Data Mining Competition Practices

Kele Xu

Data Mining Competition Practices

Methods and Cases

Tsinghua University Press

Springer

Kele Xu
National University of Defense Technology
Changsha, Hunan, China

ISBN 978-981-95-3445-6   ISBN 978-981-95-3446-3   (eBook)
https://doi.org/10.1007/978-981-95-3446-3

Jointly published with Tsinghua University Press
The print edition is not for sale in Mainland China. Customers from Mainland China please order the
print book from: Tsinghua University Press.
ISBN of Co-Publisher's edition: 978-7-302-65846-7

Translation from the Chinese Simplified language edition: "数据挖掘竞赛实战:方法与案例" by Kele
Xu, © Tsinghua University Press 2024. Published by Tsinghua University Press. All Rights Reserved.

The original submitted manuscript has been translated into English. The translation was done using
artificial intelligence. A subsequent revision was performed by the author(s) to further refine the work and
to ensure that the translation is appropriate concerning content and scientific correctness. It may, however,
read stylistically different from a conventional translation.

This Springer imprint is published by the registered company Springer Nature Singapore Pte Ltd.
The registered company address is: 152 Beach Road, #21-01/04 Gateway East, Singapore 189721,
Singapore

If disposing of this product, please recycle the paper.

Preface

Objectives of This Book

This book aims to provide readers with a clear implementation process for data mining competition solutions and explain the key details within. In addition to providing the necessary theoretical knowledge, it also offers plug-and-play code. By reading this book, readers will understand how to design a solution for a data mining competition, clarify various details and specific implementation methods within the solution, and learn how to continuously refine and optimize the solution. The book also provides specific practical cases to help readers master and strengthen the aforementioned content. Data mining competitions offer datasets that are close to real-world scenarios, making this book a great choice for those who wish to learn data mining techniques through practical experience.

At the same time, this book can also serve as a reference book, providing various methods and techniques for different scenarios (including structured data, natural language processing, computer vision, video understanding, and reinforcement learning) from data input to obtaining final results. These practical methods and techniques can help readers achieve significant improvements in their datasets, and they can be used not only in data mining competitions but also in research and actual business applications.

Who Should Read This Book

Whether you want to achieve better results in data mining competitions, improve your data mining skills, or enhance model performance in actual business applications, this book will be a great choice. The target audience for this book includes, but is not limited to, the following individuals:

- Data mining competition enthusiasts.
- University students majoring in artificial intelligence-related fields.

- Professionals working in the field of artificial intelligence.
- Readers interested in artificial intelligence.

It should be noted that due to space limitations, this book will not explain data mining knowledge points from the very beginning. Although the author tries to present the content of the book in a step by step, progressively deeper manner, ideally, the expected readers of this book should have a certain foundation in machine learning, deep learning, and reinforcement learning, as well as some experience in using Python.

If readers are familiar with the following content, it indicates that they have roughly the corresponding foundation.

- Machine Learning: Able to distinguish between supervised and unsupervised learning, understand the differences between training set, validation set, and test set, and the concept of overfitting.
- Deep Learning: Understand forward propagation and backpropagation, commonly used activation functions in neural networks, and the basic principles of stochastic gradient descent.
- Reinforcement Learning: Understand the basic concepts of Markov properties, decision environments, and environmental rewards, and common reinforcement learning algorithms such as DQN, A2C, and PPO.
- Python: Know how to execute py files in the terminal, how to use Jupyter Notebook for interactive writing and running code, and have used common Python packages related to data mining, such as Numpy and pandas.

This book focuses on how to choose the appropriate technology based on actual data scenarios and how to use these technologies in a more optimal way to achieve better results on specific datasets, rather than spending a lot of space introducing the principles of these technologies. For example, this book will not delve into the algorithmic principles of gradient boosting decision trees (GBDTs). Instead, it focuses on discussing the scenarios in which GBDT is suitable, the key hyperparameters of GBDT, and how to tune these hyperparameters more efficiently.

Book Code Description

The code[1] in the book is uniformly presented in a font different from regular text and is distinguished by a shadowed background. Key information in the code is described through comments or text.

---

[1] The source codes of the cases in some chapters of this book can be found in the corresponding chapters at https://github.com/DataBountyHunter/DataMiningCompetitionInAction.

Writing Team Members

This book is edited by Kele Xu, who, in addition to being responsible for writing Chapters 1–3, also organized the work of the entire writing team. Chapter 4 is handled by Hengwei Dai, and Chapter 5 is completed by Yanbo Wang and Sheng Chen together. Chapters 6–9 are handled by Xiaochen Cai, and the final Chapters 10 and 11 are handled by Shiyu Huang.

Acknowledgments

During the compilation of this book, we were fortunate to receive valuable support and help from many friends and colleagues.

First and foremost, special thanks go to He Yucheng, Gao Zhifeng, Liu Yuzhong, Bao Mengjiao, Fang Xi, Yan Kuo, and others (in no particular order). They provided abundant material for this book and participated in the review of its content. Their professional contributions are an indispensable part of the completion of this book. Additionally, I would like to thank Ms. Wang Qiuyang from Tsinghua University Press, who provided professional guidance and advice throughout the publication process. Finally, I would like to thank all the readers of this book, and your support is our greatest motivation. I hope this book can provide value to you, and I also look forward to receiving your suggestions and feedback.

Errata and Support

Due to the author's limited capabilities, this book may inevitably contain oversights and inaccuracies. I sincerely request readers to provide criticism and corrections.

Changsha, China   Kele Xu

Contents

Chapter 1
Introduction to Data Mining Competitions

This chapter will introduce the development history of data mining competitions and their significance in practice applications, competition platforms, characteristics of various competitions, and tools commonly used in competitions. Through the study of this chapter, readers will gain an in-depth understanding of the basic concepts and core elements of data mining competitions and learn how to effectively participate in them.

1.1 The Development of Data Mining Competitions

The development history of artificial intelligence competitions can be traced back to the 1990s (or even earlier). Some important moments in their development process are as follows.

- 1997: The first KDD Cup (International Knowledge Discovery and Data Mining Cup Competition) was held. This event is an international top-level event in the field of data mining research, organized by the Special Interest Group on Knowledge Discovery and Data Mining of the ACM (Association for Computing Machinery). It is known as the most influential event in the field of data mining, where KDD stands for knowledge discovery and data mining.
- 2006: The American video streaming company Netflix launched the Netflix Prize competition with a million-dollar prize. The goal of the competition was to improve the accuracy of Netflix's recommendation system to help users better discover content they like. This competition attracted many professionals to engage in research work in the field of recommendation systems and brought this technology from academia world into the business world.
- 2010: The Kaggle platform was established, designed specifically for developers and data scientists, providing services for hosting machine learning competitions, hosting databases, and writing and sharing code. Today, Kaggle has developed into

© Tsinghua University Press 2026

K. Xu, *Data Mining Competition Practices*,

https://doi.org/10.1007/978-981-95-3446-3_1

a key platform in the field of machine learning competitions. In that same year, the ImageNet Large-Scale Visual Recognition Challenge (ILSVRC) was launched, requiring participants to use machine learning techniques to classify large-scale image data, which greatly promoted the development of deep learning.

With the continuous development of artificial intelligence technology, the scale of the KDD Cup competition has been expanding, and the types of competitions have become increasingly diverse.

Table 1.1 shows the events of the regular track of the KDD Cup over the past decade. From it, we can see that the prize pool amount and the number of participating teams have increased.

Figure 1.1 shows the number of new registered users on the Kaggle platform over the years. It can be seen that the number of users on the Kaggle platform has

Table 1.1　Event situation of the regular track of the KDD Cup over the years

Year	Topic	Prize pool amount/USD	Participating teams
2022	ESCI challenge for improving product search	21,000	273
	Spatial dynamic wind power forecasting challenge	35,000	2490
2021	Multi-dataset time series anomaly detection	3500	614
	Large-scale challenge for machine learning on graphs	**Data missing**	More than 500
	City Brain Challenge	20,500	1156
2020	(Track 1) Multimodalities recall for E-commerce platform	17,500	1433
	(Track 2) Debiasing for E-commerce platform	17,500	1895
2019	Context-aware travel mode recommendation problem	39,000	More than 2800
2018	Fresh air prediction	36,500	More than 4000
2017	Highway tollgates traffic flow, time and volume prediction	25,000	3582
2016	Measuring the impact of research institutions	20,000	More than 500
2015	Predicting course drop on MOOC platform	20,000	821
2014	Predicting excitement at DonorsChoose.org	2000	472
2013	(Track 1) Predicting user following behavior in tencent Weibo	8000	656
	(Track 2) Predicting click-through rate of ads in tencent Weibo	8000	163

number of newly registered users

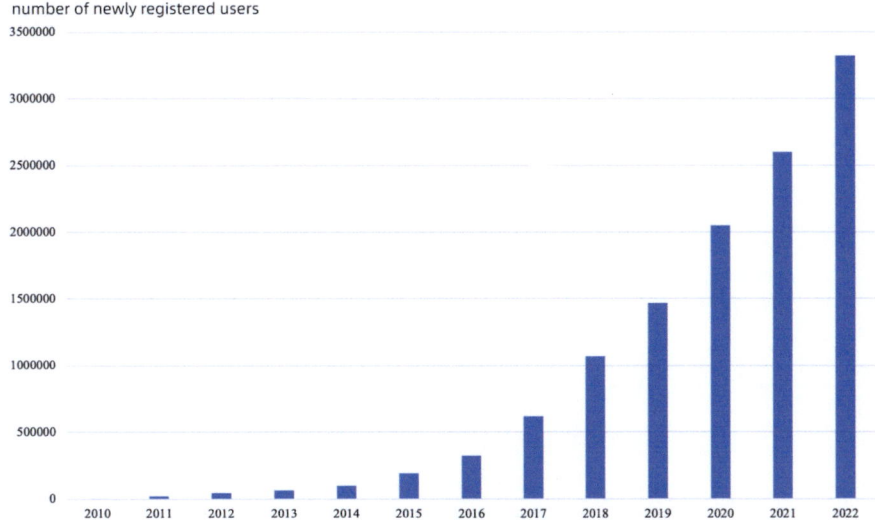

Fig. 1.1 Number of new registered users on the Kaggle platform over the years

grown very rapidly, with the number of new registered users increasing each year. According to statistics, the cumulative number of users exceeded 10 million by 2022.

Figure 1.2 shows the proportion of various competition types held on the Kaggle platform in 2022. It can be seen that structured data competitions are still the main type of data mining competitions, followed by computer vision competitions and natural language processing competitions. Reinforcement learning, as an emerging popular field in recent years, has also seen related competitions.

1.2 Significance of Data Mining Competitions

1. Significance for Organizers

The organizers of artificial intelligence competitions usually include enterprises, research institutions, industry associations, government departments, etc. Through the form of artificial intelligence competitions, they can gain the following benefits:

(1) Promote machine learning technology and applications.
(2) Attract and cultivate machine learning talent.
(3) Evaluate and compare different artificial intelligence methods and technologies.
(4) Help related industries obtain better artificial intelligence solutions.
(5) Increase the visibility and influence of the company or organization.

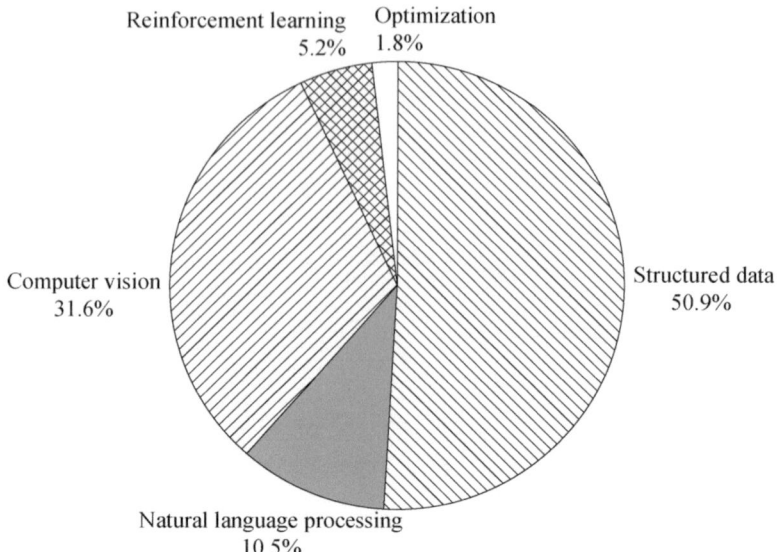

Fig. 1.2 Proportion of various competition types held on the Kaggle platform in 2022

2.   Significance for Participants

From the perspective of participants, participating in data mining competitions can bring many benefits.

(1)   Increase skills and experience in data mining

Participating in data mining competitions allows participants to enhance their skills and experience in real-world data mining tasks. These competitions provide real data and problems, allowing participants to practice the knowledge they have learned. At the same time, data mining competitions allow participants to test their algorithms in a competitive environment and receive feedback. This can help participants improve their skills more effectively.

(2)   Enrich your resume and increase job opportunities

Participating in data mining competitions and achieving good results can add an excellent project to a participant's resume. This can demonstrate the participant's practical abilities to recruiters, helping them stand out when job hunting. If a participant is interested in a particular field but lacks relevant project experience, participating in corresponding data mining competitions can open new opportunities in their career. Many data mining competitions offer a fast track to recruitment for participants who achieve good results.

(3)   Gain professional networking opportunities

Participating in data mining competitions allows you to connect with other professionals who may become your future colleagues or partners. These networking opportunities can bring many opportunities and benefits to your career. Platforms like Kaggle internationally and Tianchi domestically provide convenient discussion areas where you can exchange experiences, share techniques, meet like-minded friends, and collaborate with them.

(4)   Obtain prizes and rewards

Many data mining competitions offer cash prizes, some even reaching millions of dollars. Of course, the higher the prize, the more participants there are, and the more intense the competition.

(5)   Expand your influence

Participating in data mining competitions and achieving good results can bring more attention to your work and achievements. This is very helpful for expanding your personal influence and career development.

(6)   Help solve real-world problems

Participating in data mining competitions can help you solve real problems, which are often actual issues faced by certain companies (such as internet companies), certain fields (such as healthcare), or even society as a whole. Helping to solve these real-world problems is also a great contribution.

1.3   Introduction to Competition Platforms

1.   Kaggle

Kaggle is the largest community of data scientists and machine learning developers today, with a unique position in the industry.[1] Kaggle offers a large number of machine learning competitions (Fig. 1.3 shows some Kaggle competition projects), allowing data scientists and developers to use machine learning techniques to solve various problems and collaborate and compete with other community members. At the same time, Kaggle provides a full set of services for its community, the most distinctive of which is the Kaggle Kernels code-sharing tool.

2.   CodaLab

CodaLab is an open-source research platform designed to support research related to machine learning and artificial intelligence.[2] It provides a web interface that facilitates the ability of researchers and data scientists to upload and manage data, run experiments, and collaborate in teams. In this community, users can share worksheets and participate in competitions. Additionally, users can participate in existing competitions or host new ones. Figure 1.4 shows some competitions on CodaLab.

---

[1] The Kaggle link is https://www.kaggle.com.

[2] The CodaLab link is https://codalab.lisn.upsaclay.fr/.

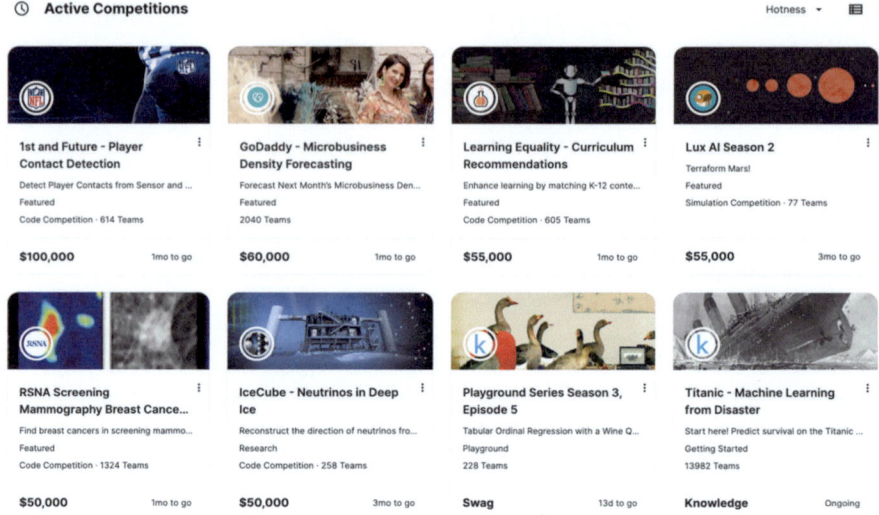

Fig. 1.3 Some Kaggle competition projects

Fig. 1.4 Some data mining competitions on the CodaLab platform

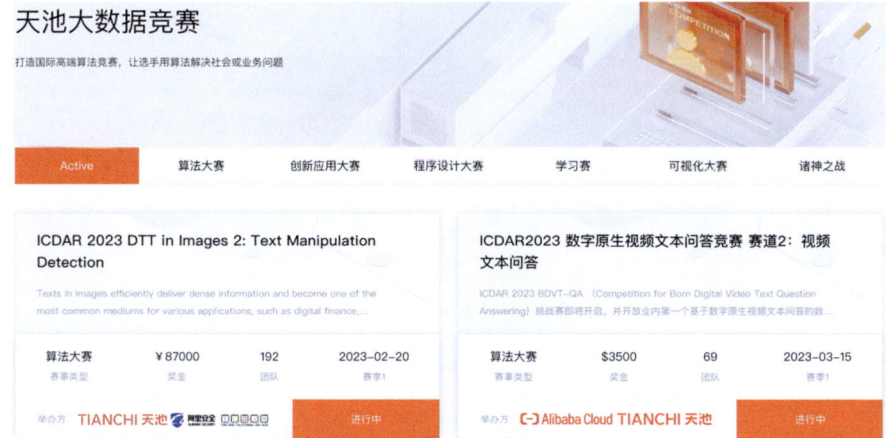

Fig. 1.5   Tianchi competition platform

3.   Tianchi

Tianchi (see Fig. 1.5) is a data mining competition platform founded by Alibaba.[3]
Similar to Kaggle, it also provides a code execution environment and a leaderboard service for participants. Additionally, the platform offers many tutorials
and knowledge sharing related to the competitions, which are very beneficial for
beginners.

4.   DataFountain

DataFountain (see Fig. 1.6) is another platform in China with a relatively rich variety
and number of competitions.[4] The most distinctive competition on this platform is
the Big Data and Computational Intelligence Contest held annually by the China
Computer Federation (CCF).

5.   Other Competition Platforms

- AIcrowd.[5]
- DataCastle.[6]
- Heywhale Community.[7]
- Biendata.
- Huawei Cloud.[8]

---

[3] The Tianchi link is https://tianchi.aliyun.com/competition/gameList/activeList.

[4] The DataFountain competition platform link is https://www.datafountain.cn.

[5] The AIcrowd link is https://www.aicrowd.com/.

[6] The DataCastle link is https://www.datacastle.cn/index.html.

[7] The Heywhale Community link is https://www.heywhale.com/home.

[8] The Huawei Cloud link is https://competition.huaweicloud.com/competitions.

Fig. 1.6   DataFountain competition platform homepage

1.4   Characteristics of Various Competitions

1.   Structured Data Competitions

These competitions provide structured data, mainly focusing on the predictive ability for numerical or categorical variables, such as predicting house prices or user purchasing behavior. Structured data refers to data in tabular form, where each row corresponds to a data sample and each column corresponds to a feature.

These competitions have the following characteristics.

(1)   Structured data is usually complex, requiring tedious preprocessing steps, with potential challenges such as a large number of missing values, high data noise, and long-tail distributions.
(2)   Feature engineering has a significant impact on the final results of these competitions.
(3)   Deep learning models are generally not the optimal choice in these competitions, so participants need to have a deep understanding of various machine learning and data mining techniques and the ability to find and construct optimal model structures.

2.   Natural Language Processing Competitions

These competitions mainly focus on the computer's ability to understand and process human language, such as text classification, sentiment analysis, and machine translation. Participants need to train models that can understand and process natural language. Due to the rapid development of natural language processing technology in recent years, the problem-solving approaches in these competitions have also changed significantly. Before 2016, the most widely used and best-performing models were the bag-of-words model and the TF-IDF model. From 2016 to 2018, word embedding models replaced the bag-of-words and TF-IDF models, and since 2019, the application of various pre-trained models has become increasingly widespread.

3. Computer Vision Competitions

These competitions mainly focus on the computer's ability to process image and video data, involving classification, segmentation, and detection of image and video data. Participants need to train models that can extract useful information from images (usually deep learning models).

These competitions have the following characteristics:

(1) The training cost of visual models is usually high, so high-performance GPU servers are important for these competitions.
(2) Data augmentation techniques and fine-tuning are important techniques for participants to improve competition results.
(3) Other techniques, such as pseudo-labeling, also help improve model performance.

4. Reinforcement Learning Competitions

These competitions mainly focus on the computer's ability to solve complex problems through trial-and-error learning, such as game AI and control systems. Participants use reinforcement learning algorithms to train robots or game characters to achieve optimal rewards in real or simulated environments. Participants need to construct reinforcement learning models that can make decisions based on the feedback they receive. In reinforcement learning competitions, multiple participants usually compete in a real or simulated environment, and the model with the highest final score is declared the winner.

These competitions have the following characteristics.

(1) Designing an appropriate reward function is important, especially for scenarios with sparse reward.
(2) Participants need to carefully design feature extraction models and reinforcement learning model structures, which are important for improving training efficiency and model performance.
(3) Consider combining some deep learning strategies and algorithms, such as on-policy and Monte Carlo Tree Search (MCTS).

1.5 Commonly Used Tools for Competitions

1. IDE

Figure 1.7 shows the number of participants in data mining competitions using various common IDE tools, as surveyed by Kaggle in 2022 with approximately 24,000 participants. The statistical results show that the top three IDE tools used by participants are Jupyter Notebook, VSCode, and PyCharm.

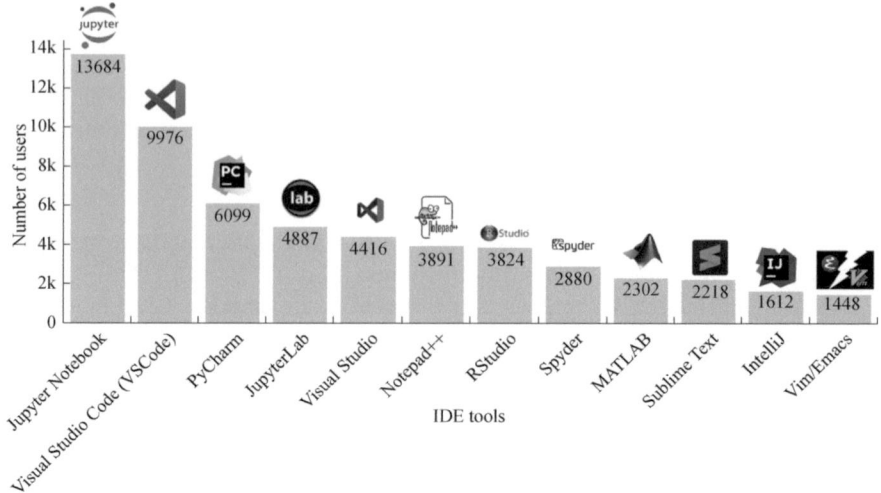

Fig. 1.7 Usage of IDE tools by Kaggle participants in 2022

(1)   Jupyter Notebook

Jupyter Notebook is a web-based application that supports an interactive computing environment, allowing users to develop and execute code, as well as write documentation and visualize results.

(2)   VS Code

VS Code is a code editor launched by Microsoft that supports multiple programming languages and features syntax highlighting, auto-completion, debugging, and other functionalities. It also has many installable plugins that can help users write code more conveniently.

(3)   PyCharm

PyCharm is an integrated development environment (IDE) for Python developed by JetBrains, integrating numerous practical features including code auto-completion, syntax highlighting, debugging, and version control, which can help you develop Python applications more efficiently.

2.   Machine Learning Libraries

Figure 1.8 shows the number of users using various common machine learning libraries according to the 2022 Kaggle official survey. The statistical results show that the number of people using Scikit-learn is far ahead. In addition, the commonly used machine learning frameworks for implementing gradient boosting decision trees, including XGBoost, LightGBM, and CatBoost, have a large number of users.

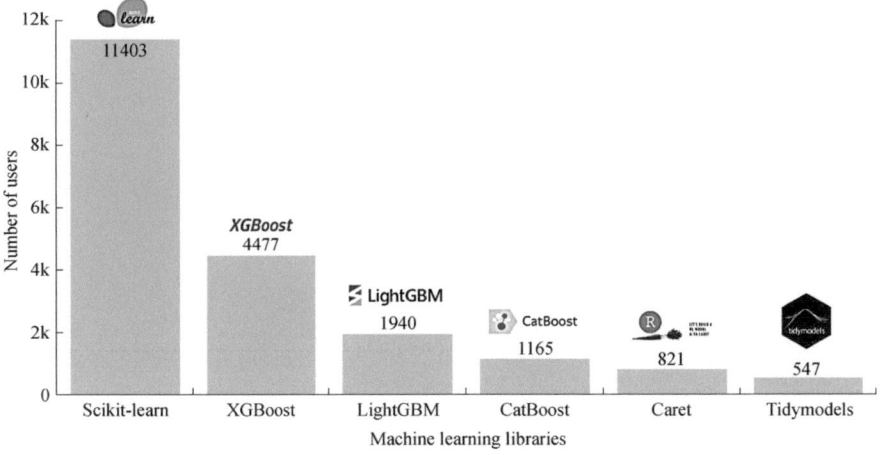

Fig. 1.8 Usage of machine learning libraries by 2022 Kaggle participants

(1)   Scikit-learn

Scikit-learn is a widely used open-source machine learning library that supports many common machine learning algorithms, including but not limited to regression, classification, clustering, and dimensionality reduction. It also provides users with a simple and consistent interface to use these algorithm models.

(2)   XGBoost, LightGBM, CatBoost

XGBoost, LightGBM, and CatBoost are three commonly used frameworks for implementing gradient boosting decision trees, especially suitable for structured data scenarios. Each has its own advantages and characteristics: XGBoost was the earliest to originate and was widely used in early structured data mining competitions; LightGBM emphasizes being lightweight, achieving good results in a shorter time; CatBoost has special optimizations for categorical variables, which can yield better results in certain scenarios.

3.   Deep Learning Libraries

Figure 1.9 shows the number of users using various common deep learning libraries according to the 2022 Kaggle official survey. The statistical results show that Tensor-Flow, Keras, and PyTorch remain the three most popular mainstream deep learning frameworks.

(1)   TensorFlow

TensorFlow is an open-source deep learning framework by Google, allowing users to quickly design deep learning networks without writing low-level CUDA or C++ code. TensorFlow excels in code simplicity, execution efficiency, and deployment convenience.

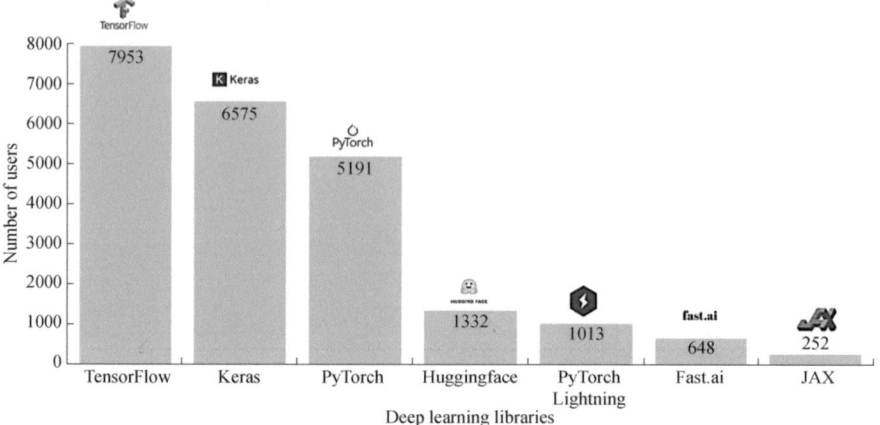

Fig. 1.9 Usage of deep learning libraries by 2022 Kaggle participants

(2) Keras

Compared to TensorFlow, Keras is a higher-level API. Keras is very friendly to beginners in deep learning, allowing users to build an understandable network model with just a few lines of code through its API. Currently, Keras has been integrated into TensorFlow, and all Keras tasks can be completed on TensorFlow.keras API.

(3) PyTorch

PyTorch is an open-source deep learning framework by Facebook's AI Research Lab. Due to its flexible interface, ease of use, and speed, it has become the mainstream deep learning framework in academia, with most deep learning model papers, except those by Google, using PyTorch for experiments.

4. Visualization Libraries

Figure 1.10 shows the number of users using visualization libraries according to the 2022 Kaggle participants' statistics. It can be seen that the top three most popular visualization libraries are Matplotlib, Seaborn, and Plotly.

(1) Matplotlib

Matplotlib is the veteran of Python data visualization libraries and remains the most widely used plotting library in the Python community, capable of generating various required graphs with just a few lines of code.

(2) Seaborn

Seaborn is a higher-level encapsulation based on Matplotlib, providing a more concise syntax than Matplotlib, making it easier for users to get started and produce more aesthetically pleasing graphs.

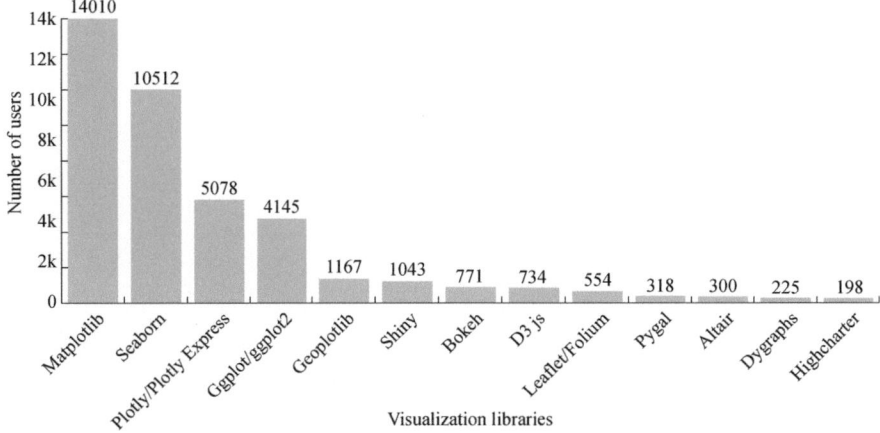

Fig. 1.10 Usage of visualization libraries by 2022 Kaggle participants

(3) Plotly

Plotly is a platform that provides data visualization capabilities, excelling in interactive plotting, offering a complete range of chart types, and enabling online sharing and open-source contributions.

Chapter 2
Structured Data: Theoretical Part

The typical process of structured data competitions includes exploratory data analysis (EDA), data preprocessing, feature engineering (including feature construction and feature selection), modeling (including model selection and model hyperparameter optimization), and ensemble learning. The structured data modeling process is illustrated in Fig. 2.1.

2.1 Exploratory Data Analysis

Exploratory data analysis is a method that uses visualization techniques to analyze data. As Scott Berinato noted in his book *Good Charts*, "A good visualization can communicate the nature and potential impact of information and ideas more powerfully than any other form of communication."

In machine learning, proper data preprocessing and selection of suitable features play a crucial role in subsequent model training. EDA can help us discover certain patterns and trends in the data and validate certain hypotheses using statistical descriptive information and graphical representations, thereby guiding our approach to data preprocessing and feature engineering.

In EDA, key information to focus on includes missing values, outliers, data distribution, correlation between variables, and correlation between variables and labels. Tools like Matplotlib and Seaborn can be used to manually conduct EDA, but a more convenient method is to use automated EDA tools such as D-Tale, pandas profiling, sweetviz, and AutoViz. Here, we take AutoViz as an example, using the Iris dataset to demonstrate how to perform EDA with just a few lines of code, as illustrated below.

© Tsinghua University Press 2026
K. Xu, *Data Mining Competition Practices*,
https://doi.org/10.1007/978-981-95-3446-3_2

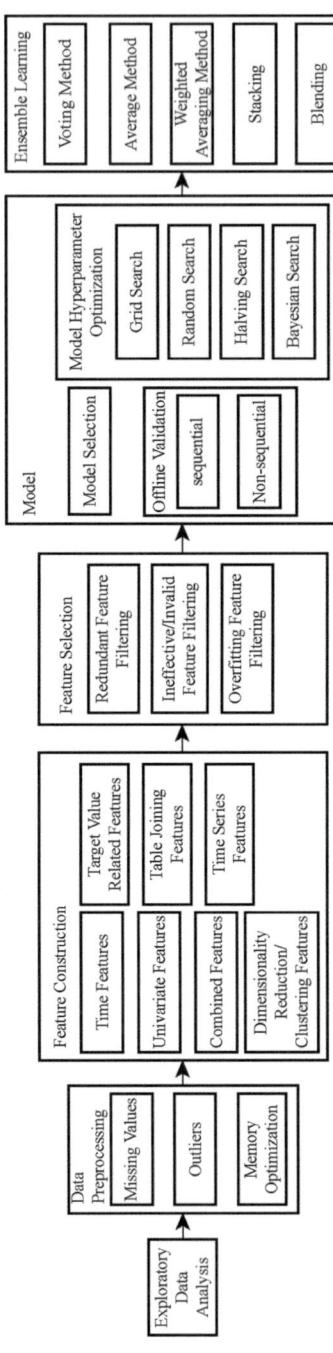

Fig. 2.1 Structured data modeling process

```
From autoviz.AutoViz_Class import AutoViz_Class
   autoviz = AutoViz_Class()
   dft =autoviz.AutoViz(
   filename="/content/Iris.csv",   # Load the dataset, note the
difference with dft
       sep=",",                     # Set the dataset delimiter,
default is comma
       depVar="Species",           # Set the label column
       dfte=None, # Pass a pandas.DataFrame, if filename is set,
this should be None
       header=0,
        verbose=0, # Optional 0, 1, or 2, set the form of graph
saving
        lowess=False,# Whether to enable lowess regression,
suitable for small datasets, not recommended for datasets over
100,000 rows
       chart_format="svg",         # Set the graph saving format
       max_rows_analyzed=150000, # Set the number of rows to be
analyzed in the dataset
       max_cols_analyzed=30     # Set the number of columns to be
analyzed in the dataset
   )
```

AutoViz can compute the importance of each variable, select those with higher importance for visualization, and apply built-in heuristic algorithms to determine the optimal visualization format. In this case, AutoViz generated the visual results as shown in Figs. 2.2, 2.3, 2.4, 2.5, 2.6 and 2.7.

Figure 2.2 shows four scatter plots, each displaying the relationship between a feature and the Iris species (label). Among them:

- Species: Represents the Iris species (Iris-setosa, Iris-versicolor, Iris-virginica).
- SepalLengthCm: Represents sepal length.
- SepalWidthCm: Represents sepal width.
- PetalLengthCm: Represents petal length.
- PetalWidthCm: Represents petal width.

Figure 2.3 shows the pair plot of the Iris dataset, which comprises four continuous variables (sepal length, sepal width, petal length, and petal width). Each scatter plot illustrates the relationship between two variables, and each Iris species (Iris-setosa, Iris-versicolor, Iris-virginica) is marked with a different color.

Figure 2.4 shows the distribution of four features (petal width, petal length, sepal length, and sepal width) across different species of iris flowers in the dataset. These plots are known as kernel density estimate (KDE) plots, which smoothly depict the distribution shape of the data.

Figure 2.5 shows two bar charts representing the species distribution in the iris dataset: one depicting percentage distribution and the other the frequency distribution.

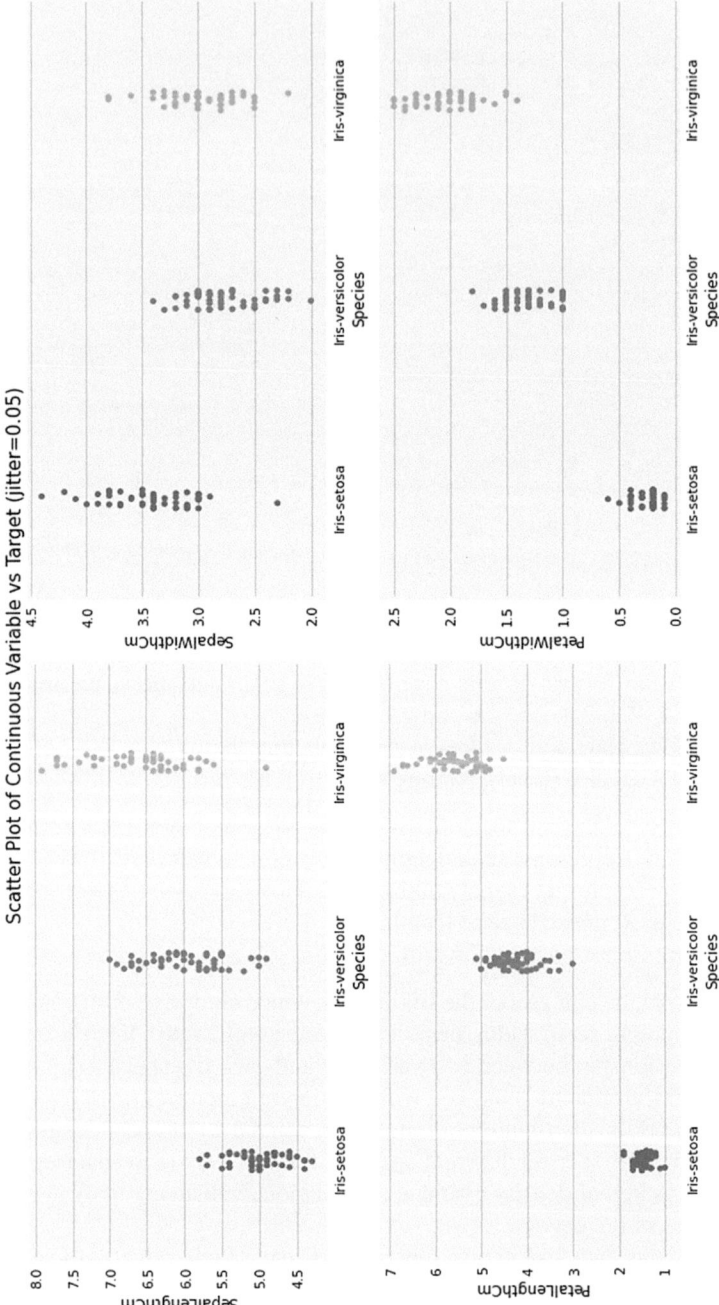

Fig. 2.2 Scatter plot (showing the relationship between continuous variables and labels)

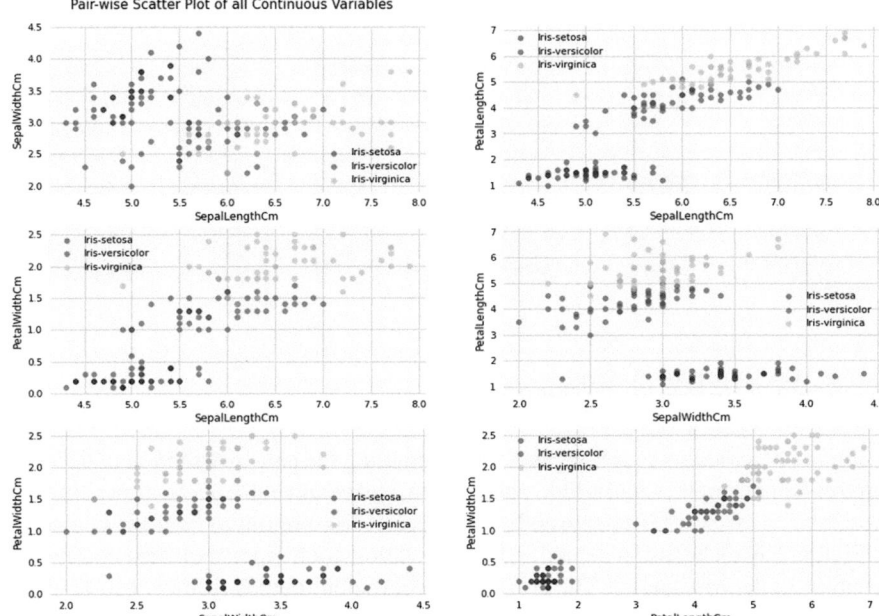

Fig. 2.3 Pair plot (used to understand the degree of correlation between continuous variables)

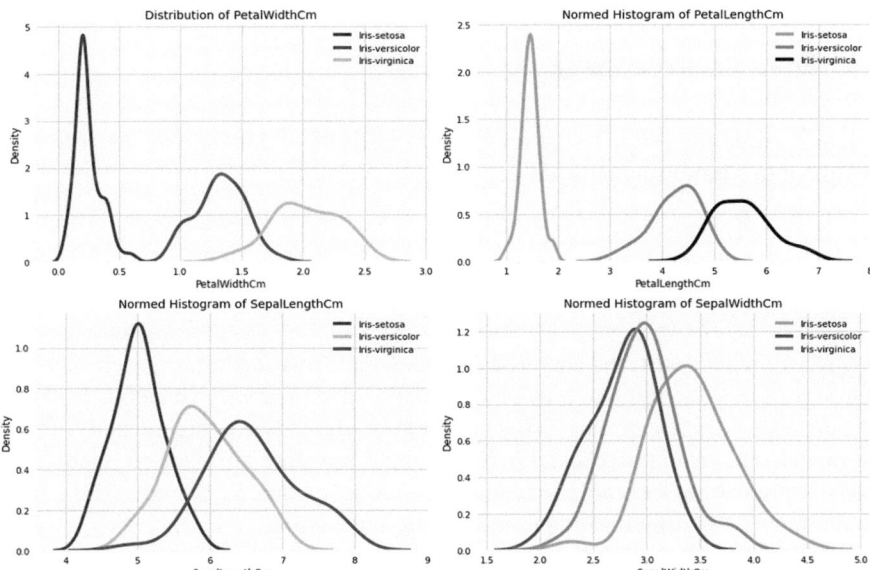

Fig. 2.4 Data distribution of continuous variables

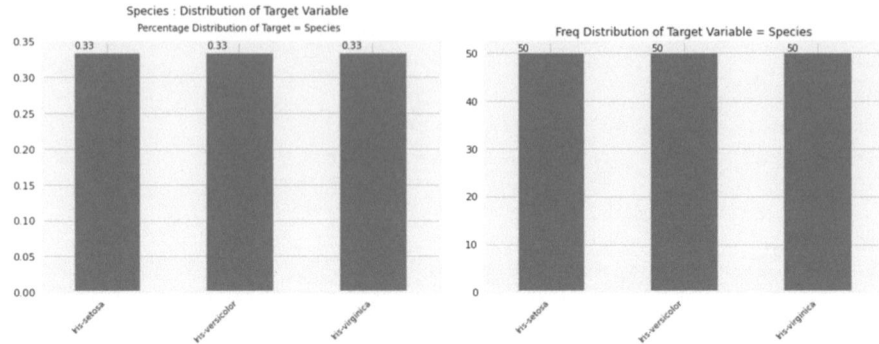

Fig. 2.5 Data distribution of labels

Figure 2.6 contains four box plots, each depicting the data distribution of different features (sepal length, sepal width, petal length, and petal width) across three different species of iris flowers (Iris-setosa, Iris-versicolor, Iris-virginica) in the iris dataset.

Figure 2.7 is a heatmap depicting the correlation coefficients between continuous variables in the iris dataset. The correlation coefficient ranges from -1 to 1, where 1 is indicating a perfect positive correlation, -1 is indicating a perfect negative correlation, and 0 is indicating no correlation.

2.2   Data Preprocessing

Data preprocessing refers to the preliminary analysis of data and its conversion of the data into a more standardized and normalized format.

2.2.1   Missing Values

Due to various reasons, the acquired data may have missing values, such as sensor failures or users unwilling to provide information due to privacy concerns. Handling missing values is trivial, but different handling methods can significantly impact the model. The main purposes of handling missing values are: first, to improve the model's prediction accuracy; second, some models (such as LR and NN) cannot handle inputs with missing values and require processing.

You can calculate the percentage of missing values in each column using the following command.

```
df.isna().sum()/len(df)*100
```

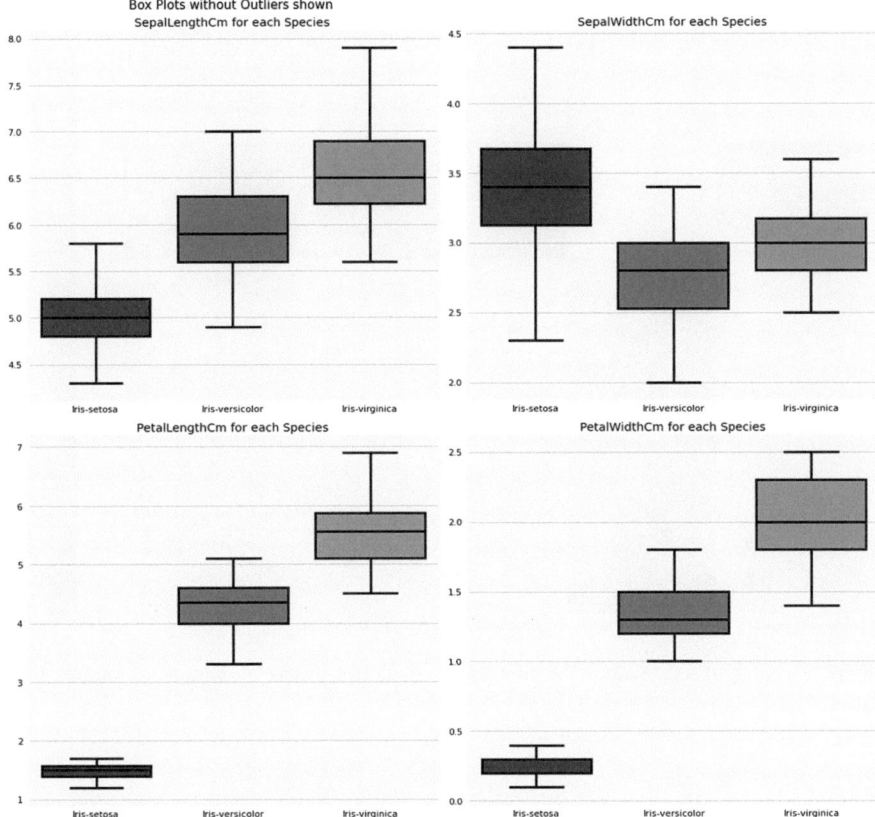

Fig. 2.6 Box plots of continuous variables under different labels

The method for handling missing values should be chosen based on the specific scenario, primarily comprising the following three methods.

1. Do Not Process

Some models (such as LightGBM) can directly handle missing values through built-in algorithms, so there is no need to process missing values in this case.

2. Missing Value Deletion Method

The missing value deletion method includes deleting features and deleting samples. For scenarios with a high proportion of missing values, this method can be attempted, as shown in the code below.

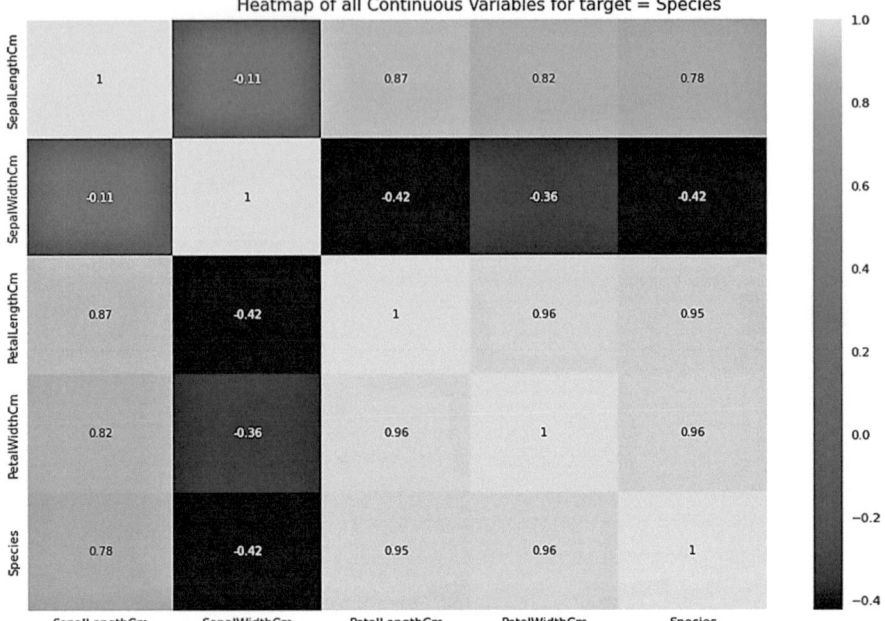

Fig. 2.7 Heatmap (used to show the correlation coefficient matrix of continuous variables and labels)

```
df.drop([5, 6], inplace=True)                    # Delete rows
df.drop(['feature_1'],  axis=1,  inplace=True)    #  Delete
columns
```

3. Use Representative Values to Fill

Using representative values to fill includes the following four methods.

(1) Fill with Values Outside the Data Range

Filling with values outside the data range can retain the original information of the missing values, as shown in the code below.

```
df['feature_2'].fillna(df['feature_2'].min()        -        1,
inplace=True)
```

(2) Fill with Statistical Values

Filling with statistical values (such as mean or median) can make the data more conform to a normal distribution, as shown in the code below.

```
# Fill with mean
df['feature_3'].fillna(df['feature_3'].mean(),
inplace=True)
# Fill with median
df['feature_3'].fillna(df['feature_3'].median(),
inplace=True)
```

(3)   Fill with Adjacent Values

If there is some connection between adjacent samples (such as representing two consecutive sampling results of a sensor), adjacent values can be employed for filling, as shown in the code below.

```
df['feature_4'].fillna(method='ffill', inplace=True) # Fill
with the previous valid observation
df['feature_4'].fillna(method='bfill', inplace=True) # Fill
with the next valid observation
```

(4)   Fill with Predicted Values

When the data volume is large enough and the column to be filled is relatively easy to fit, the method of filling with predicted values can be used. This method results in more accurate filled data and can minimize the impact of human data analysis, as shown in the code below.

```
from sklearn.linear_model import LinearRegression

# Select non-empty rows of the feature_6 column from the
dataframe as training data
train = df.loc[df['feature_6'].notnull()]
# Select rows where the 'feature_6' column is null from the
dataframe as test data
test = df.loc[df['feature_6'].isnull()]
# Set the target column as 'feature_6'
target = 'feature_6'
# Select all features except the target column as training
features
used_features = [x for x in train.columns if x != target]
# Create a linear regression model
lr = LinearRegression()
# Fit the linear regression model with training data
lr.fit(train[used_features], train[target])
# Use the linear regression model to predict on test data
pred = lr.predict(test[used_features])
# Fill the predicted results into the feature_6 column of the
original dataframe
```

```
df.loc[df['feature_6'].isnull(), 'feature_6'] = pred
```

2.2.2   Outliers

Outliers refer to values that do not conform to business logic, deviate from the normal range, or show significant differences from other samples. Properly handling outliers can prevent them from interfering with models and avoid incorrect conclusions.

1.   Outlier Identification

Outlier identification means detecting abnormal data samples through data analysis. Common methods include those based on business logic and data visualization.

(1)   Based on Business Logic

Rules can be set to determine if data is abnormal. For example, a person's height and weight cannot be negative, and postal codes must consist of 6 digits.

(2)   Based on Data Visualization

The box plot is a simple yet effective visualization method for identifying outliers. As shown in Fig. 2.8, data points outside the upper and lower limits can be considered outliers, where:

$$\text{Upper limit} = Q3 + 1.5 \times \text{interquartile range (IQR)}$$

$$\text{Lower limit} = Q1 - 1.5 \times \text{interquartile range (IQR)}$$

Note: $Q3$ and $Q1$ represent the 75th and 25th percentiles (quartiles), respectively. The code for creating a boxplot is as follows.

```
import seaborn as sns
sns.boxplot(data=data)
```

2.   Outlier Handling

Identified outliers should be handled according to specific contexts. If an outlier results from errors in data collection or recording, it needs to be corrected; if it represents a real phenomenon, it should be included in model considerations. For the former, outlier handling methods can refer to missing value treatments, such as deletion, filling with representative values, or filling with predicted values.

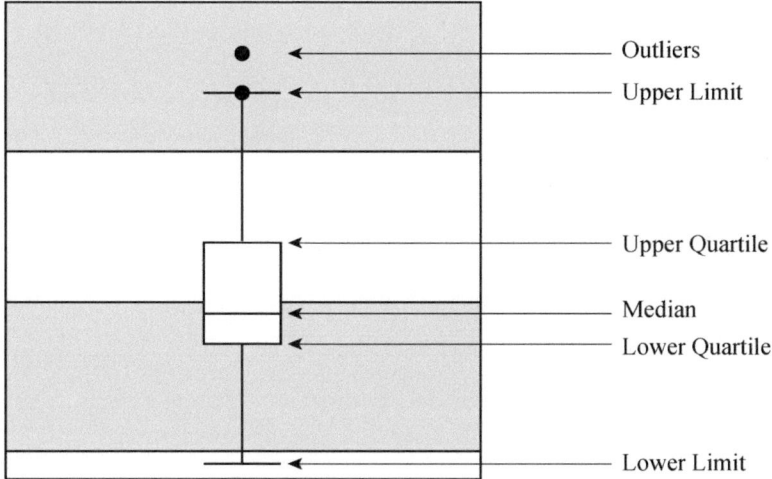

Fig. 2.8 Box plot

2.2.3 *Memory Optimization*

During data processing, errors due to insufficient memory (e.g., running out of memory) may occur. When device memory cannot be changed, memory optimization can be attempted.

1. Memory Reclamation

Delete unnecessary variables and release memory occupied by destroyed objects:

```
del df
import gc
gc.collect()
```

2. Using Data Types with Smaller Memory Footprint

The following code applies to Pandas DataFrames. For each numerical feature, convert it to the smallest data type that can accommodate its value range (original code: https://www.kaggle.com/code/gemartin/load-data-reduce-memory-usage/notebook).

```python
import numpy as np

def reduce_mem_usage(df):
    """
    Iterate through all columns of a DataFrame and modify data
    types to reduce memory usage.
    Parameters:
    df: pandas.DataFrame, the DataFrame to be optimized for
    memory.
    Returns:
    Optimized DataFrame
    """
    # Calculate initial memory usage
    start_mem = df.memory_usage().sum() / 1024**2
    print(f'Memory usage of dataframe is {:.2f} MB.format(start_
    mem))

    # Iterate over each column in the DataFrame
    for col in df.columns:
        col_type = df[col].dtype

        # If the column data type is not object
        if col_type != object:
            c_min = df[col].min()
            c_max = df[col].max()

            # If the column data type is integer
            if str(col_type)[:3] == 'int':
                # Check if convertible to int8
                if c_min > np.iinfo(np.int8).min and c_max <
np.iinfo(np.int8).max:
                    df[col] = df[col].astype(np.int8)
                # Check if convertible to int16
                elif c_min > np.iinfo(np.int16).min and c_max <
np.iinfo(np.int16).max:
                    df[col] = df[col].astype(np.int16)
                # Check if convertible to int32
                elif c_min > np.iinfo(np.int32).min and c_max <
np.iinfo(np.int32).max:
                    df[col] = df[col].astype(np.int32)
                # Default to int64
                else:
                    df[col] = df[col].astype(np.int64)
            # If the column data type is floating-point
            else:
                # Check if convertible to float16
                if c_min > np.finfo(np.float16).min and c_max <
np.finfo(np.float16).max:
                    df[col] = df[col].astype(np.float16)
                # Check if convertible to float32
                elif c_min > np.finfo(np.float32).min and c_max <
np.finfo(np.float32).max:
                    df[col] = df[col].astype(np.float32)
```

```
                # Default to float64
                else:
                    df[col] = df[col].astype(np.float64)
        else:
            # If the data type of a column is object, convert it to
category type
            df[col] = df[col].astype('category')

    # Calculate optimized memory usage
    end_mem = df.memory_usage().sum() / 1024**2
    print(f'Memory usage after optimization is: {end_mem:.2f}
MB'.format(end_mem))
    print(f'Decreased by {:.1f}%'.format(100 * (start_mem - end_
mem) / start_mem))

    return df

# Optimize memory usage of DataFrame df using reduce_mem_usage
df = reduce_mem_usage(df)
```

Note: After this operation, if further transformations are applied to these columns, ensure that the transformed data does not exceed the range of data type; otherwise, it will impact subsequent modeling.

2.3 Feature Construction

Feature construction is a very important part of machine learning tasks. Its purpose is to transform raw data into features that can be understood by machine learning algorithms, in order to better describe the data and thereby enhance the model's generalization ability and predictive performance.

2.3.1 Time Features

In actual data, the time column is usually represented in the form of timestamps, such as "2019-11-20 14:30:00." Timestamps cannot be directly used as model inputs, and it is necessary to extract more fine-grained time attributes from them. Features that can be extracted from timestamps include year, month, day, hour, calendar week, day of the week, whether it is a weekend, quarter, whether it is the start of the month, whether it is the end of the month, and whether it is a holiday.

```
import pandas as pd

def get_time_feature(df, col):
    """
```

```
    Extract time features from the time column and add them to the
DataFrame.

    Parameters:
    df: pandas.DataFrame, the DataFrame containing the time
column
    col: str, the name of the time column

    Returns:
    The DataFrame with added time features
    """

    prefix = col + "_"

    # Convert the time column to datetime type
    df[col] = pd.to_datetime(df[col])

    # Extract the year and add it as a new column
    df[prefix + 'year'] = df[col].dt.year

    # Extract the month and add it as a new column
    df[prefix +'month'] = df[col].dt.month

    # Extract the day and add it as a new column
    df[prefix + 'day'] = df[col].dt.day

    # Extract the hour and add it as a new column
    df[prefix + 'hour'] = df[col].dt.hour

   # Extract the week number in the year and add it as a new column
    df[prefix + 'weekofyear'] = df[col].dt.weekofyear

    # Extract the day of the week and add it as a new column
    df[prefix + 'dayofweek'] = df[col].dt.dayofweek

    # Determine whether the date is a weekend and add it as a new
column
    df[prefix + 'is_wknd'] = df[col].dt.dayofweek // 5

    # Extract the quarter and add it as a new column
    df[prefix + 'quarter'] = df[col].dt.quarter

    # Determine whether the date is the start of the month and add
it as a new column
        df[prefix + 'is_month_start'] = df[col].dt.is_month_
start.astype(int)

    # Determine whether the date is the end of the month and add
it as a new column
        df[prefix + 'is_month_end'] = df[col].dt.is_month_
end.astype(int)

    # Determine whether the date is a holiday and add it as a new
column
    df[prefix + 'is_holiday'] = df[col].apply(lambda x: 1 if is_
holiday(x) else 0)

    return df
```

```
# Use the get_time_feature function to extract time features
from the 'time' column and add them to the DataFrame df
df = get_time_feature(df, "time")
```

2.3.2 Univariate Features

Univariate feature extraction refers to the process of extracting features by performing certain transformations on a single variable. Variables are usually categorized into continuous and discrete variables, each having common methods for univariate feature extraction.

1. Continuous Variables

Continuous variables are those that can take any value within a certain range, such as real numbers, rational numbers, or special irrational numbers, like height, weight, and temperature. Given the wide range of values that continuous variables can take, univariate feature extraction is crucial for handling continuous variables.

(1) Binning Features

Binning features assign continuous variables to discrete bins, which helps improve the stability of the model because binned features are less susceptible to outliers.

```
df['continuous_var_bin'] = pd.cut(df['continuous_var'], 10,
labels=False)
```

(2) Box–Cox Transformation

The Box–Cox transformation is a generalized power transformation method that can significantly improve the normality, symmetry, and homoscedasticity of data, which can be beneficial for model training in certain cases. The common log transformation is a special form of the Box–Cox transformation. In regression problems, it is common to apply a log transformation to the labels, fit the model to the transformed values, and then perform an inverse operation on the results after prediction. In practice, $\log 1p$ is often used instead of log, and its inverse operation is the expm1 function.

```
import scipy.stats as st
df['continuous_var_box-cox'], _ = st.boxcox(df['continuous_
var'])
df['continuous_var_log1p'] = np.log1p(df['continuous_var'])
```

(3) Ranking Features

In some cases, the relative relationship between values is more important than the values themselves, and ranking features can effectively express this information.

```
df['continuous_var_rank'] = df['continuous_var'].rank()
```

2. Discrete Variables

Discrete variables are those that can only take a finite number of values, such as nominal, ordinal, or binary variables, like gender, occupation, and education level.

(1) Count Features

Count features refer to calculating the frequency of discrete variables in the dataset.

```
df['discrete_var_count']        =        df.groupby(['discrete_
var'])['discrete_var'].transform('count')
```

(2) Ranking Features

Some discrete variables inherently have ranking information, such as grades A, B, and C. In this case, it is necessary to extract the ranking features of these variables.

```
df['discrete_var_rank']    =    df['discrete_var'].map({'A':1,
'B':2, 'C':3})
```

(3) LabelEncoder

LabelEncoder maps the categories of a variable to integers.

```
from sklearn.preprocessing import LabelEncoder
le = LabelEncoder()
```

```
df['discrete_var_LabelEncoder']              =              le.fit_
transform(df['discrete_var'])
```

(4) One-hot

One-hot encoding transforms a single variable into multiple columns (the number of columns is equal to the number of categories of the variable), with each column representing whether the sample is a certain value of the feature. Each column's value is either composed of 0 or 1, where 1 indicates yes, and 0 indicates no.

```
pd.get_dummies(df['discrete_var'], prefix='discrete_var')
```

2.3.3 Combined Features

Unlike univariate features, combined features refer to constructing new features by combining multiple features. Combined features are divided into three categories: discrete feature × continuous feature, discrete feature × discrete feature, and continuous feature × continuous feature.

1. Discrete Feature × Continuous Feature

Use discrete features for grouping and aggregation to calculate the statistical information of continuous features within each group. Statistical information includes maximum, minimum, median, mean, variance, sum, kurtosis, skewness, ranking features, and percentages.

```
# Calculate maximum, minimum, median, mean, variance, sum
df.groupby('discrete_var1').agg({'continuous_var1':
['max', 'min',
'median', 'mean', 'std', 'sum']})
# Calculate skewness and kurtosis
from scipy.stats import skew, kurtosis
df.groupby('discrete_var1').agg({'continuous_var1': [skew,
kurtosis]})
# Ranking features
df['continuous_var1-rank'] = df.groupby('discrete_var1')
['continuous_var1'].rank()
# Percentage
df['percentage'] = 100 * df['continuous_var1'] /
```

```
df.groupby('discrete_var1')['continuous_
var1'].transform('sum')
```

2. Discrete Feature × Discrete Feature

Use one of the discrete features for grouping and aggregation to calculate the number of unique categories of another discrete feature within the group.

```
# Number of unique categories within the group
df.groupby('discrete_var1').agg({'discrete_var2':
['nunique']})
```

Concatenate two discrete features to form a single discrete feature, and then use the feature extraction method for discrete univariate features.

```
# Concatenation of discrete features
df['discrete_var1-var2'] = df['discrete_var1'].astype(str)
+ '-' +
df['discrete_var2'].astype(str)
```

3. Continuous Feature × Continuous Feature

Use binary operations to calculate two columns of continuous features, including addition, subtraction, multiplication, division, and modulus.

```
# Binary operations
df['continuous_var1_add_continuous_var2'] = df['continuous_
var1'] +
df['continuous_var2']
df['continuous_var1_sub_continuous_var2'] = df['continuous_
var1']   -
df['continuous_var2']
df['continuous_var1_mul_continuous_var2'] = df['continuous_
var1']   *
df['continuous_var2']
df['continuous_var1_div_continuous_var2'] = df['continuous_
var1']   /
df['continuous_var2']
df['continuous_var1_mod_continuous_var2'] = df['continuous_
var1']   %
df['continuous_var2']
```

Calculate statistical information for multiple columns of features, including maximum, minimum, median, mean, variance, and sum.

```
# Calculate statistical values for multiple columns of features
df['continuous_var1_var2_max'] = df[['continuous_var1' ,
'continuous_var2']].max(axis=1)
df['continuous_var1_var2_min'] = df[['continuous_var1' ,
'continuous_var2']].min(axis=1)
df['continuous_var1_var2_median'] = df[['continuous_var1'
,
'continuous_var2']].median(axis=1)
df['continuous_var1_var2_mean'] = df[['continuous_var1' ,
'continuous_var2']].mean(axis=1)
df['continuous_var1_var2_std'] = df[['continuous_var1' ,
'continuous_var2']].std(axis=1)
df['continuous_var1_var2_sum'] = df[['continuous_var1' ,
'continuous_var2']].sum(axis=1)
```

2.3.4   Dimensionality Reduction/Clustering Features

1. Dimensionality Reduction Features

Dimensionality reduction features refer to features constructed by transforming the original data using various dimensionality reduction techniques. The correlation between dimensionality reduction features is relatively low, and the information expressed by a single feature is relatively dense. Typical dimensionality reduction techniques include principal component analysis (PCA), independent component analysis (ICA), singular value decomposition (SVD), Gaussian random projection (GRP), sparse random projection (SRP), and non-negative matrix factorization (NMF).

```
from sklearn.decomposition import PCA, FastICA, TruncatedSVD,
NMF
from sklearn.random_projection import GaussianRandomProjec-
tion, SparseRandomProjection

n_comp = 3 # Set the number of features after dimensionality
reduction

# Use PCA for dimensionality reduction
pca = PCA(n_components=n_comp, random_state=42) # Create a PCA
object, setting the number of features after dimensionality
reduction and the random seed
```

```
pca_df  =  pd.DataFrame(pca.fit_transform(df))  #  Perform
dimensionality reduction on the data and store the result
in a DataFrame

# Use ICA for dimensionality reduction
ica = FastICA(n_components=n_comp, random_state=42) # Create
an ICA object, setting the number of features after dimension-
ality reduction and the random seed
ica_df  =  pd.DataFrame(ica.fit_transform(df))  #  Perform
dimensionality reduction on the data and store the result
in a DataFrame

# Use tSVD for dimensionality reduction
tsvd = TruncatedSVD(n_components=n_comp, random_state=42) #
Create a tSVD object, setting the number of features after
dimensionality reduction and the random seed
tsvd_df = pd.DataFrame(tsvd.fit_transform(df))  # Perform
dimensionality reduction on the data and store the result in a
DataFrame

# Use GRP for dimensionality reduction
grp       =       GaussianRandomProjection(n_components=n_comp,
eps=0.1,  random_state=42)  #  Create  a  GRP  object,  setting
the number of features after dimensionality reduction, the eps
value, and the random seed
grp_df  =  pd.DataFrame(grp.fit_transform(df))  #  Perform
dimensionality reduction on the data and store the result
in a DataFrame

# Use SRP for dimensionality reduction
srp  =  SparseRandomProjection(n_components=n_comp,  dense_
output=True, random_state=42) # Create an SRP object, setting
the number of features after dimensionality reduction, the
dense_output parameter, and the random seed
srp_df  =  pd.DataFrame(srp.fit_transform(df))  #  Perform
dimensionality reduction on the data and store the result
in a DataFrame

# Use NMF for dimensionality reduction
nmf  =  NMF(n_components=n_comp,  init='nndsvdar',  random_
state=42) # Create an NMF object, setting the number of features
after dimensionality reduction, the initialization method,
and the random seed
nmf_df  =  pd.DataFrame(nmf.fit_transform(df))  #  Perform
dimensionality reduction on the data and store the result
in a DataFrame
```

2. Clustering Features

Clustering features refer to using unsupervised clustering algorithms to cluster the data and using the category to which the sample belongs as a feature. Typical clustering algorithms include k-means clustering algorithm and spectral clustering algorithm.

```
from sklearn.cluster import KMeans, SpectralClustering

# k-means clustering algorithm
kms = KMeans(n_clusters=3, random_state=1).fit(df)
df['kmeans_Cluster'] = kms.labels_

# Spectral clustering algorithm
sc = SpectralClustering().fit(df)
df['sc_Cluster'] = sc.labels_
```

2.3.5 Target Value Related Features

1. Target Encoding

The main idea of target encoding is to encode the target based on the impact of
categorical variables on the target value. A common method is to calculate the mean of
the target value for each category. Target encoding is a powerful feature construction
method, but care must be taken to avoid target leakage during use. To avoid target
leakage, cross-validation is usually used.

```
from sklearn.model_selection import KFold
def fe_target_encoding(train, test, key, label, k = 5):
    """
    This function is used for target encoding.

    Parameters:
    train: Training dataset
    test: Testing dataset
    key: Name of the feature to be encoded
    label: Name of the target variable
    k: Number of splits for KFold, default is 5

    Returns:
    oof_train: Encoded result of the training set
    oof_test: Encoded result of the testing set
    """

    # Create two zero arrays to store the encoded data
        oof_train,  oof_test  =  np.zeros(train.shape[0]),
    np.zeros(test.shape[0])

    # Use K - fold cross - validation to create indices for the
    training and validation sets
    skf = KFold(n_splits = k).split(train)

    # Iterate over each fold
    for i, (train_idx, valid_idx) in enumerate(skf):
```

```
        # Get the training and validation sets according to the
    indices
        df_train = train[key + [label]].loc[train_idx]
        df_valid = train[key].loc[valid_idx]

    # Group the training set and calculate the mean of the target
    variable for each group
        df_map = df_train.groupby(key)[[label]].agg('mean')

        # Map the mean values of the training set to the validation
    set
        oof_train[valid_idx] = df_valid.merge(df_map, on = key,
    how = 'left')[label].values

        # Map the mean values of the training set to the testing set
    and handle missing values
        oof_test += test[key].merge(df_map, on = key, how =
    'left')[label].fillna(-1).values / k

    # Return the encoded training and testing sets
    return oof_train, oof_test
```

2.   Gradient Boosting Decision Tree Features

Gradient boosting decision tree (GBDT) is an ensemble learning algorithm that uses boosting technology. In each iteration step, it constructs a new weak learner (decision tree) to compensate for the shortcomings of the current model. The prediction result of GBDT is the sum of the prediction results of multiple sequential decision trees.

The feature construction method of GBDT is as follows: Suppose we have trained a GBDT model with N trees, and each tree's leaf nodes are numbered (see Fig. 2.9). When we input a sample into the GBDT for prediction, the sample will fall into a leaf node in each tree. These leaf node numbers of these trees can be used as features, resulting in N features.

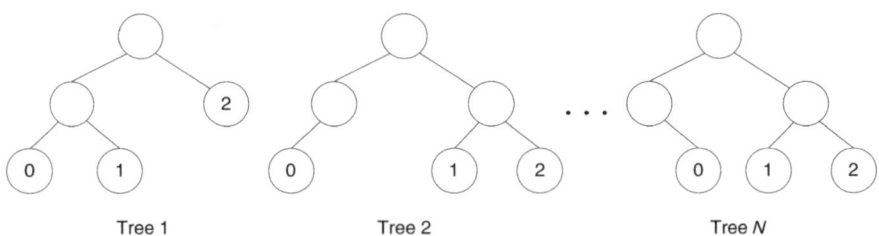

Fig. 2.9 Numbering the leaf nodes of each tree in GBDT

```python
import lightgbm as lgb
def fe_gbdt(train, test, used, category, label):
    """
    This function is used to perform LightGBM encoding on the
input features.

    Parameters:
    train: Training dataset
    test: Testing dataset
    used: Names of the feature columns to be used
    category: Names of the categorical feature columns
    label: Name of the target variable

    Returns:
    train_lgb_feature: Encoded result of the training set
    test_lgb_feature: Encoded result of the testing set
    """

    # Define the parameters of the LightGBM model
    params = {
        'num_leaves': 41,
        'min_child_weight': 0.03,
        'feature_fraction': 0.3,
        'bagging_fraction': 0.4,
        'min_data_in_leaf': 96,
        'objective': 'binary',
        'max_depth': -1,
        'learning_rate': 0.01,
        "boosting_type": "gbdt",
        "bagging_seed": 11,
        "metric": 'auc',
        "verbosity": -1,
        'reg_alpha': 0.4,
        'reg_lambda': 0.6,
        'random_state': 47,
        'num_threads': -1
    }
    # Define the number of iterations
    N_round = 30

    # Build the dataset for training
    trn_data = lgb.Dataset(train[used], label=train[label],
categorical_feature=category)

    # Train the model
    clf = lgb.train(params, trn_data, num_boost_round=N_round,
valid_sets=[trn_data], verbose_eval=10)

    # Use the model to make predictions on the training and testing
sets, and use the results as new features
    train_lgb_feature = pd.DataFrame(clf.predict(train[used],
pred_leaf=True))
    test_lgb_feature = pd.DataFrame(clf.predict(test[used],
pred_leaf=True))
```

```
# Name the new features
tree_feas = ["gbdt_" + str(i) for i in range(1, N_round + 1)]
train_lgb_feature.columns = tree_feas
test_lgb_feature.columns = tree_feas

# Return the encoded training and testing sets
return train_lgb_feature, test_lgb_feature
```

2.3.6   Table Joining Features

In table-type data mining competitions, there are often multiple data tables. To fully utilize the information from these tables, it is necessary to concatenate multiple tables. Here, we discuss the common scenarios of one-to-one and one-to-many table join.

1.   One-to-One Table Join

In the scenario of one-to-one table join, you only need to concatenate two tables using a join key. In the following example, Table 2.1 contains User_id, Gender, Occupation, and corresponding label information, while Table 2.2 contains more information about the users. The second table can be concatenated to the first table using the User_id, and the concatenated result is shown in Table 2.3.

Table 2.1   User basic information

User_id	Gender	Occupation	Label
000014b8ec0ce8ad7c20f56915fc3a9f	1	2	0
0003f283dfacd7100bba76d876cf94da	1	4	0
0015f0bf7222ad1b2a96612d752552c3	1	2	0
00251f8fa9f3e014339d90e4dba1affd	1	2	0
00284cf15ae27d1ddf4f93922cd7bcb5	1	2	0

Table 2.2   User additional information

User_id	Education_level	Marital_status	Household_type
000014b8ec0ce8ad7c20f56915fc3a9f	3	1	2
0003f283dfacd7100bba76d876cf94da	4	1	2
0015f0bf7222ad1b2a96612d752552c3	4	2	2
00251f8fa9f3e014339d90e4dba1affd	4	3	1
00284cf15ae27d1ddf4f93922cd7bcb5	4	3	1

Table 2.3 One-to-one table join result

User_id	Gender	Occupation	Education_level	Marital_status	Household_type	Label
000014b8ec0ce8ad7c20f56915fc3a9f	1	2	3	1	2	0
0003f283dfacd77100bba76d876cf94da	1	4	4	1	2	0
0015f0bf7222ad1b2a96612d752552c3	1	2	4	2	2	0
0025f8fa9f3e014339d90e4dba1affd	1	2	4	3	1	0
00284cf15ae27d1ddf4f93922cd7bcb5	1	2	4	3	1	0

2 One-to-Many Table Join

In the scenario of a one-to-many table join, it is necessary to first extract features from the secondary table and then concatenate them to the main table. In the following anti-fraud scenario, the goal is to predict whether a user defaults. Each user in the main table (see Table 2.4) is a sample, containing basic information (such as gender and occupation) and whether the user defaults. The secondary table (see Table 2.5) represents the user's bank card transaction records, where each user may have multiple transaction records, and thus the main table and the secondary table have a one-to-many relationship.

Aggregated statistics are the methods of combining datasets into a single result, often used to combine related data from multiple data sources into one dataset for further analysis.

(1) Aggregated Statistics

The most common method of passing subtable information to the main table is to aggregate and calculate statistical information from the subtable, such as calculating the average transaction amount and total transaction amount in the user's bank card transaction records.

```
continue_ops = ['max', 'min', 'median', 'mean', 'std', 'sum']
discrete_ops = ['nunique']

temp = bank_train.groupby(' user_id ').agg({"transaction_
type ': discrete_ops, "transaction_amount ': stat_ops, '
salary_income_mark ': discrete_ops, 'month': discrete_ops}).
reset_index()

# Rename
rename_cols = []
for item in temp.columns:
    if item[1] != ":
       rename_cols.append('_'.join(item))
    else:
       rename_cols.append(".join(item))
temp.columns = rename_cols

train = train.merge(temp, on = ' user_id ', how = 'left')
```

(2) Constructing Subtable Meta-features

Constructing subtable meta-features is a more efficient and thorough way of utilizing subtable information for feature construction, as is shown in Fig. 2.10.

Taking the data from Tables 2.4 and 2.5 as an example, the steps to construct subtable meta-features are as follows.

(1) Transfer the user's label from the main table to the subtable, so that each transaction record in the subtable has a label.

Table 2.4 User information table

User_id	Gender	Occupation	Education_level	Marital_status	Household_type	Label
000014b8ec0ce8ad7c20f56915fc3a9f	1	2	3	1	2	0
0003f283dfacd71100bba76d876cf94da	1	4	4	1	2	0
0015f0bf7222ad1b2a96612d752552c3	1	2	4	2	2	0
00251f8fa9f3e014339d90e4dba1affd	1	2	4	3	1	0
00284cf15ae27d1ddf4f93922cd7bcb5	1	2	4	3	1	0

Table 2.5   User bank card transaction record table

User_id	Transaction_type	Transaction_amount	Salary_income_mark	Month
000014b8ec0ce8ad7c20f56915fc3a9f	1	38.134741	0	1
000014b8ec0ce8ad7c20f56915fc3a9f	1	40.189051	0	1
000014b8ec0ce8ad7c20f56915fc3a9f	2	42.143743	1	1
000014b8ec0ce8ad7c20f56915fc3a9f	1	40.125193	0	1
000014b8ec0ce8ad7c20f56915fc3a9f	1	39.317371	0	1

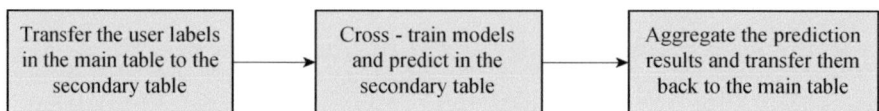

Fig. 2.10   Process of constructing subtable meta-features

(2) Randomly split the subtable (assuming five folds), cross-validate five times, using four folds for training each time, and predict on the remaining fold and the test set to obtain prediction results. The prediction results of the training set are the concatenation of the five predictions, and the prediction results of the test set are the average of the five predictions.

(3) Aggregate the prediction results to obtain statistical values of the prediction results, such as maximum, minimum, mean, and variance, and pass them back to the main table.

```
# Transfer the user labels in the main table to the secondary
table
bank_train = bank_train.merge(train[['user_id', 'label']],
on='user_id', how='left')

# Perform five - fold cross - validation, cross - train and
predict in the secondary table
Id = ['user_id']
target = 'label'
used = ['transaction_type', 'transaction_amount','salary_
income_mark','month']

bank_train['predict'] = -1
test_predict = np.zeros(bank_test.shape[0])

folds = StratifiedKFold(n_splits=5, shuffle=True, random_
state=42)

for n_fold, (trn_idx, val_idx) in enumerate(folds.split(bank_
train[used], bank_train[target])):
    print(f'n_fold: {n_fold + 1}')
```

```
        clf = LogisticRegression(random_state=0)
                clf.fit(bank_train.loc[trn_idx][used],    bank_
train.loc[trn_idx][target])
        bank_train.loc[val_idx, 'predict']  = clf.predict_
proba(bank_train.loc[val_idx][used])[:, 1]
    test_predict += clf.predict_proba(bank_test[used])[:, 1]
/ folds.n_splits

bank_test['predict'] = test_predict

# Aggregate the prediction results and transfer them back to the
main table
stat_ops = ['max','min','median','mean','std', 'count']

temp = bank_train.groupby('user_id').agg({'predict': stat_
ops}).reset_index()
temp.columns = ['_'.join(x) if x[1] != " else ".join(x) for x
in list(temp.columns)]
train = train.merge(temp, on='user_id', how='left')

temp  =  bank_test.groupby('user_id').agg({'predict': stat_
ops}).reset_index()
temp.columns = ['_'.join(x) if x[1] != " else ".join(x) for x
in list(temp.columns)]
test = test.merge(temp, on='user_id', how='left')
```

(3) Construct Auxiliary Table Meta-features Using Time Series Models

When the auxiliary table contains time series information, you can also consider using time series models to construct auxiliary table meta-features. The process of constructing auxiliary table meta-features using time series models is shown in Fig. 2.11.

Below, using the data from Tables 2.4 and 2.5 as examples, the steps to construct auxiliary table meta-features are as follows.

(1) Sort the auxiliary table by the join key (user_id) and the time column (month).
(2) Perform aggregate statistics on the auxiliary table by the join key (user_id) and the time column (month), calculating the mean of each feature for each user per month.
(3) Organize the data into the input format for the time series model.

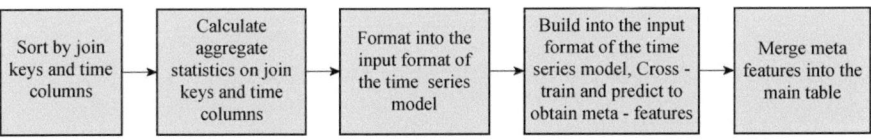

Fig. 2.11 Process of constructing auxiliary table meta-features using time series models

(4) Build the time series model, cross-train the model and predict to obtain time series meta-features. The prediction results of the training set are the concatenation of multi-layer cross-prediction results, and the prediction results of the test set are the average of multiple prediction results.

(5) Merge the meta-features into the main table.

```python
# Sort the secondary table by the join key and the time column
bank_train = bank_train.sort_values(by=['user_id','month'])
bank_test = bank_test.sort_values(by=['user_id','month'])

# Perform aggregated statistics on the secondary table by the
join key and the time column
num_aggregations = {
    'transaction_type': ['mean'],
    'transaction_amount': ['mean'],
    'salary_income_mark': ['mean']
}

bank_train_g            =            bank_train.groupby(['user_
id','month']).agg(num_aggregations)
bank_test_g             =            bank_test.groupby(['user_
id','month']).agg(num_aggregations)
bank_train_g.columns = ['transaction_type', 'transaction_
amount','salary_income_mark']
bank_test_g.columns = ['transaction_type', 'transaction_
amount','salary_income_mark']

# Organize the data into the input format of the time - series
model: number of samples × number of time steps × number of
feature columns
train_x = np.array(bank_train_g.to_xarray().to_array())
train_x = train_x.swapaxes(0, 1).swapaxes(1, 2)

test_x = np.array(bank_test_g.to_xarray().to_array())
test_x = test_x.swapaxes(0, 1).swapaxes(1, 2)

train_x[np.isnan(train_x)] = -9
test_x[np.isnan(test_x)] = -9

train_y = np.array(train[target])

# Build the time - series model
def build_model(time_step, n_features):
    model = Sequential()
    model.add(GRU(8, input_shape=(time_step, n_features))) #
unit: # of neurons in each LSTM cell? input_shape=(time_step,
n_features)
    model.add(Dense(1, activation='sigmoid'))
    return model

class IntervalEvaluation(Callback):
    def __init__(self, validation_data=(), interval=1):
        super(Callback, self).__init__()
```

```
        self.interval = interval
        self.X_val, self.y_val = validation_data

    def on_epoch_end(self, epoch, logs={}):
        if epoch % self.interval == (self.interval - 1):
            y_pred = self.model.predict(self.X_val, verbose=0)[:,
0]

            score = roc_auc_score(self.y_val, y_pred)
            print('roc score', score)

# Cross - train the model and make predictions to obtain time -
series meta - features
folds = StratifiedKFold(n_splits=5, shuffle=True, random_
state=777)
oof_preds = np.zeros(train_x.shape[0])
sub_preds = np.zeros(test_x.shape[0])

for n_fold, (trn_idx, val_idx) in enumerate(folds.split(train_
x, train_y)):
    trn_x, val_x = train_x[trn_idx], train_x[val_idx]
    trn_y, val_y = train_y[trn_idx], train_y[val_idx]
    ival = IntervalEvaluation(validation_data=(val_x, val_y),
interval=5)

    model = build_model(trn_x.shape[1], trn_x.shape[2])

        model.compile(loss='binary_crossentropy', opti-
mizer=Adam(decay=0.0005))

    model.fit(trn_x, trn_y,
        validation_data=(val_x, val_y),
        epochs=5, batch_size=5000,
        class_weight={0: 1, 1: 10},
        callbacks=[ival], verbose=5)

    oof_preds[val_idx] = model.predict(val_x)[:, 0]
    sub_preds += model.predict(test_x)[:, 0] / folds.n_splits

    print('Fold %2d AUC : %.6f' % (n_fold + 1, roc_auc_score(val_
y, oof_preds[val_idx])))

    del model, trn_x, trn_y, val_x, val_y
    gc.collect()

pos_score_train = pd.DataFrame({'user_id': train['user_
id'], 'gru_feature': oof_preds})
pos_score_test = pd.DataFrame({'user_id': test['user_id'],
'gru_feature': sub_preds})

# Merge the meta - features into the main table
train    =    train.merge(pos_score_train,    on='user_id',
how='left')
test = test.merge(pos_score_test, on='user_id', how='left')
```

2.3.7 *Time Series Features*

Time series features refer to extracting features based on the time sequence for data with time information. Extracting time series features can help us better understand the time-varying patterns of data and has important practical application value.

When there is time information in data samples, time series-related features can be extracted from the data. Figure 2.12 shows a data table with a time column.

It should be noted that when extracting time series features, the data needs to be sorted by the time column first.

```
df = df.sort_values(by='time').reset_index(drop=True)
```

1. Lag Features

Lag features mean that after aggregating a certain categorical variable, the information corresponding to the previous N (or next N) samples under the same categorical variable is obtained. For example, for the data in Fig. 2.12, we can obtain

	User ID	time	Consumption Amount
0	17	2020-01-01 00:30:04	17.0
1	38	2020-01-01 01:14:13	87.0
2	36	2020-01-01 02:48:01	12.0
3	36	2020-01-01 02:50:20	4.0
4	31	2020-01-01 03:14:22	6.0
...	...	...	...
995	49	2020-01-30 21:02:24	81.0
996	55	2020-01-30 21:26:39	37.0
997	9	2020-01-30 21:43:22	36.0
998	19	2020-01-30 22:58:34	41.0
999	15	2020-01-30 23:01:06	86.0

Fig. 2.12 Time series feature example data

the consumption amount corresponding to the previous purchase record of the same user.

```
key = 'user_id'
val = 'consumption_amount'
step = 1
name = f'{key}_{val}_lag_{step}'
df[name]    =    df.groupby(key)[val].transform(lambda    x:
x.shift(step))
```

2. Diff Features

Diff features are a further transformation based on lag features. After sorting by time, a certain column is aggregated, and then the difference between the current sample of another column and the previous N (or next N) samples is calculated. For example, after sorting the data in Fig. 2.12 by time, we calculate the difference between the consumption amount of the user's current consumption record and that of the previous consumption record.

```
key = 'user_id'
val = 'consumption_amount'
step = 1
name = f'{key}_{val}_diff_{step}'
lag_val = df.groupby(key)[val].shift(step).values
origin_val = df.groupby(key)[val].shift(0).values
df[name] = lag_val - origin_val
```

3. Statistical Features Within a Time Series Window

Statistical features within a time series window mean that after sorting by time and aggregating a certain column, records within a certain recent window range (the window can be a fixed number of observations or a fixed time range) are obtained, and the statistical features (such as mean, variance, median, maximum, and minimum) of a certain column within the window are calculated.

```
key = 'user_id'
val = 'consumption_amount'
window = 3
ops = ['mean','std','median','max','min']

for op in ops:
    name = f'{key}_{val}_rolling_{window}_{op}'
    if op =='mean':
```

```
        df[name] = df.groupby(key)[val].transform(lambda x:
x.rolling(window=window).mean())
    if op =='std':
        df[name] = df.groupby(key)[val].transform(lambda x:
x.rolling(window=window).std())
    if op =='median':
        df[name] = df.groupby(key)[val].transform(lambda x:
x.rolling(window=window).median())
    if op =='max':
        df[name] = df.groupby(key)[val].transform(lambda x:
x.rolling(window=window).max())
    if op =='min':
        df[name] = df.groupby(key)[val].transform(lambda x:
x.rolling(window=window).min())
```

It should be noted that in most data mining competitions, it is allowed to use samples with later time than the current record to construct features, and such features sometimes have significant effects. In actual business scenarios, such features may not be available when deployed online.

2.4   Feature Selection

Using too many features to train a model can bring some side effects, so it is necessary to filter these features that may bring side effects. The idea of feature filtering is mainly divided into two categories. The first category is the filtering of redundant, invalid, or inefficient features. These features can lead to inefficiency, possibly causing exhaustion of hard disk or memory, or leading to excessively long model training and prediction times. The second category is features that cause overfitting, which can lead to a decline in model performance. Different methods can be used to filter these different types of features.

2.4.1   Redundant Feature Filtering

Redundant features refer to two features that convey highly overlapping information, and at that point one of the features can be deleted by calculating the correlation coefficient to obtain the degree of information overlap between the two features.

```
def redundant_feature_filter(df, threshold=0.9):
    """
```

```
    This function is used to filter highly correlated features
in a dataset.

    Parameters:
    df: The input DataFrame dataset.
    threshold: The correlation threshold used to determine
whether to exclude a feature, with a default value of 0.9.

    Returns:
    redundant_features: A list of the names of the filtered
features.
    """
    # Calculate the correlation matrix of the input dataset
    corr = df.corr()

    # Create a boolean array with the same length as the number of
columns, initialized with all values as True
    columns = np.full((corr.shape[0],), True, dtype=bool)

    # Iterate through each element in the correlation matrix
    for i in range(corr.shape[0]):
        for j in range(i + 1, corr.shape[0]):
            # If the correlation between two features is higher than
the set threshold
            if corr.iloc[i, j] >= threshold:
                # If the feature has not been marked as False yet, mark
it as False, indicating that the feature needs to be excluded
                if columns[j]:
                    columns[j] = False

    # Get the names of the retained features
    selected_columns = list(df.columns[columns])

    # Get the names of the filtered features
    redundant_features = [x for x in df.columns if x not in
selected_columns]

    # Return the names of the filtered features
    return redundant_features
```

2.4.2 Invalid/Inefficient Features Filtering

This section introduces four common methods for filtering invalid or inefficient features: variance threshold, feature importance in linear models, feature importance in tree models, and permutation importance. These methods can be used to evaluate the features in a dataset to remove useless or inefficient features, thereby improving the prediction accuracy and interpretability of the model.

1. Varience Threshold

Variance represents the degree of dispersion of a feature. If the variance of a feature is close to or equal to 0, the model cannot learn useful information from this feature, and it will only unnecessarily increase the complexity of the model. Therefore, such features should be removed.

```python
def variance_filter(df, threshold=1e-10):
    low_var_features = list(df.columns[df.var() < threshold])
    return low_var_features
```

2. Feature Importance in Linear Models

Linear models (such as logistic regression, linear regression, ridge regression, and Lasso regression) use the weighted sum of input features to make predictions. The absolute values of these weighting coefficients can serve as indicators of feature importance.

```python
def get_lr_importance(df, used, target):
    from sklearn.linear_model import LogisticRegression
    model = LogisticRegression()
    model.fit(df[used], df[target])
    importance = model.coef_[0]
    lr_importance = pd.DataFrame(df[used].columns)
    lr_importance.columns = ['feature']
    lr_importance['importance'] = abs(importance)
    lr_importance = lr_importance.sort_values(by = 'impor-
tance', ascending = False).reset_index(drop = True)
    return lr_importance
```

3. Feature Importance in Tree Models

Tree models (such as decision trees, random forests, XGBoost, and LightGBM) use one feature at a time to split the data space. The importance of features can be characterized by the degree of a feature's participation in the construction process of a decision tree (for example the number of times a feature is used for splitting and the information gain associated with split node values).

```python
from sklearn.ensemble import RandomForestClassifier

def get_rf_importance(df, used, target):
    """
    This function is used to obtain the importance of features.

    Parameters:
```

```
        df: The input DataFrame dataset.
        used: The names of the feature columns used for training the
    model.
        target: The name of the target variable.

        Returns:
         rf_importance: A DataFrame is returned, containing each
    feature and its corresponding importance.
        """

        # Create a random forest model
        model = RandomForestClassifier()

        # Train the model
        model.fit(df[used], df[target])

        # Obtain the feature importance of the model
        importance = model.feature_importances_

        # Create a DataFrame containing the name of each feature
        rf_importance = pd.DataFrame(df[used].columns)
        rf_importance.columns = ['feature']

        # Add the importance of each feature to the DataFrame
        rf_importance['importance'] = abs(importance)

        # Sort according to the feature importance and reset the index
                     rf_importance    =    rf_importance.sort_
    values(by='importance',              ascending=False).reset_
    index(drop=True)

        # Return the DataFrame containing feature importance
    return rf_importance
```

4. Permutation Importance

Permutation importance is a method for evaluating the importance of features in a machine learning model. It measures the importance of a feature by calculating the degree of change in the model's prediction performance after randomly permuting a certain feature while keeping other features unchanged.

The higher the importance of a feature column, the greater the loss of the model's prediction accuracy will be if the order of this feature is randomly shuffled. Permutation importance calculates the feature importance precisely based on this logic. When the order of a feature column is randomly shuffled while other features and the label column remain unchanged, the resulting loss in the model's prediction accuracy is used to characterize the importance of the shuffled feature.

```
from sklearn.linear_model import LogisticRegression
from sklearn.inspection import permutation_importance

def get_permutation_importance(df, used, target):
```

```
    """
    This function is used to obtain the permutation importance
of features.

    Parameters:
    df: The input DataFrame dataset.
    used: The names of the feature columns used for training the
model.
    target: The name of the target variable.

    Returns:
    permutation_importance: A DataFrame is returned, containing
each feature and its corresponding permutation importance.
    """

    # Get the names of the feature columns used for training the
model, excluding the target column
    used = [x for x in df.columns if x != target]

    # Create and train a logistic regression model
    clf = LogisticRegression().fit(df[used], df[target])

    # Calculate the permutation importance of each feature
    result = permutation_importance(clf, df[used], df[target],
n_repeats=10, random_state=0)

    # Create a DataFrame containing the name of each feature
    permutation_importance = pd.DataFrame(df[used].columns)
    permutation_importance.columns = ['feature']

    # Add the permutation importance of each feature to the
DataFrame
                permutation_importance['importance']    =
result.importances_mean

    # Sort according to the feature permutation importance and
reset the index
        permutation_importance = permutation_importance.sort_
values(by='importance',             ascending=False).reset_
index(drop=True)

    # Return the DataFrame containing feature permutation impor-
tance
    return permutation_importance
```

2.4.3 Overfitting Feature Filtering

1. Null Importance

The null importance method can be used to assess the stability of features. It is based
on the following assumption: By shuffling the order of the labels, the importance

of the features is calculated both before and after shuffling. For a stable feature, the importance after shuffling will decrease significantly; for an unstable feature, the importance after shuffling will be close to or even greater than the feature importance of the correct label model.

```
def get_null_importance(df, used, target):
    # Define an inner function to obtain feature importance
    def get_feature_importances(df, used, target, shuffle,
seed=None):
        from sklearn.ensemble import RandomForestClassifier
        y = df[target].copy() # Get the target values
        if shuffle:
            y = df[target].copy().sample(frac=1.0) # If the shuffle
parameter is True, shuffle the target column
        model = RandomForestClassifier() # Create a random forest
classifier model
        model.fit(df[used], y) # Fit the model
        imp_df = pd.DataFrame() # Create an empty DataFrame to
store feature importance
        imp_df["feature"] = used # Add the feature name column
        imp_df["importance"] = model.feature_importances_ # Add
the feature importance column
        return imp_df

    actual_imp_df = get_feature_importances(df, used, target,
shuffle=False) # Obtain the actual feature importance

    shuffle_imp_df = pd.DataFrame() # Create an empty DataFrame
to store the shuffled importance
    nb_runs = 50 # Set the number of shuffled runs
    for i in tqdm(range(nb_runs), total=nb_runs): # Display a
progress bar for each run
        imp_df = get_feature_importances(df, used, target,
shuffle=True)
        # Calculate and obtain the feature importance for each run
        imp_df['run'] = i + 1 # Add the run number column
        shuffle_imp_df = pd.concat([shuffle_imp_df, imp_df],
axis=0) # Concatenate the feature importance of all runs into
one DataFrame

    null_imp_df = pd.DataFrame() # Create an empty DataFrame to
store the null importance
    null_imp_df['feature'] = used # Add the feature name column
    null_imp_df['importance'] = 0 # Initialize the importance
column to 0
    for feature in used: # Operate on each feature
        # Calculate the null importance of each feature: actual
importance minus the average shuffled importance
        null_imp_df.loc[null_imp_df['feature'] == feature,
'importance'] = \
        actual_imp_df.loc[actual_imp_df['feature'] == feature,
'importance'].values[0] - \
```

```
        shuffle_imp_df.loc[shuffle_imp_df['feature'] ==
feature, 'importance'].mean()

    # Sort the null importance, with features of higher importance
ranked first, and reset the index
        null_imp_df = null_imp_df.sort_values(by='importance',
ascending=False).reset_index(drop=True)

    return null_imp_df # Return the DataFrame of null importance
```

2. Adversarial Validation

Adversarial validation can filter out features with inconsistent distributions between the training set and the test set. The execution logic of adversarial validation is shown in Fig. 2.13. First, set the labels of the training set to 1 and the labels of the test set to 0, and then merge the training set and the test set. Next, use the merged dataset to perform cross-validation and prediction to obtain the AUC on the validation set. If the distribution of features in the training set and the test set is relatively consistent, the obtained AUC will be close to 0.5, as the model cannot distinguish between the training set and the test set based on the features. If the AUC is greater than a threshold (e.g., 0.6), the feature with the highest importance can be identified and removed. Repeat the cross-validation process until the AUC falls below the threshold.

```
def   adversarial_validation(train,   test,   used,   target,
threshold=0.6):
    # Import the required libraries
    from sklearn.model_selection import KFold
    from sklearn.ensemble import RandomForestClassifier
    from sklearn.metrics import roc_auc_score

    train[target] = 1 # Set the target column of the training set
to 1
    test[target] = 0 # Set the target column of the test set to 0
    train_test = pd.concat([train, test], axis=0) # Concatenate
the training set and the test set

    n_fold = 5 # Set the number of folds for K-fold cross-
validation
    folds = KFold(n_splits=n_fold, shuffle=True, random_
state=889) # Initialize KFold

    removed_features = [] # Create an empty list to store the
removed features

    while True: # Start an infinite loop, which will exit only
when the exit condition is met
        print('#' * 50)
        AUCs = [] # Create an empty list to store the AUC scores of
each fold
```

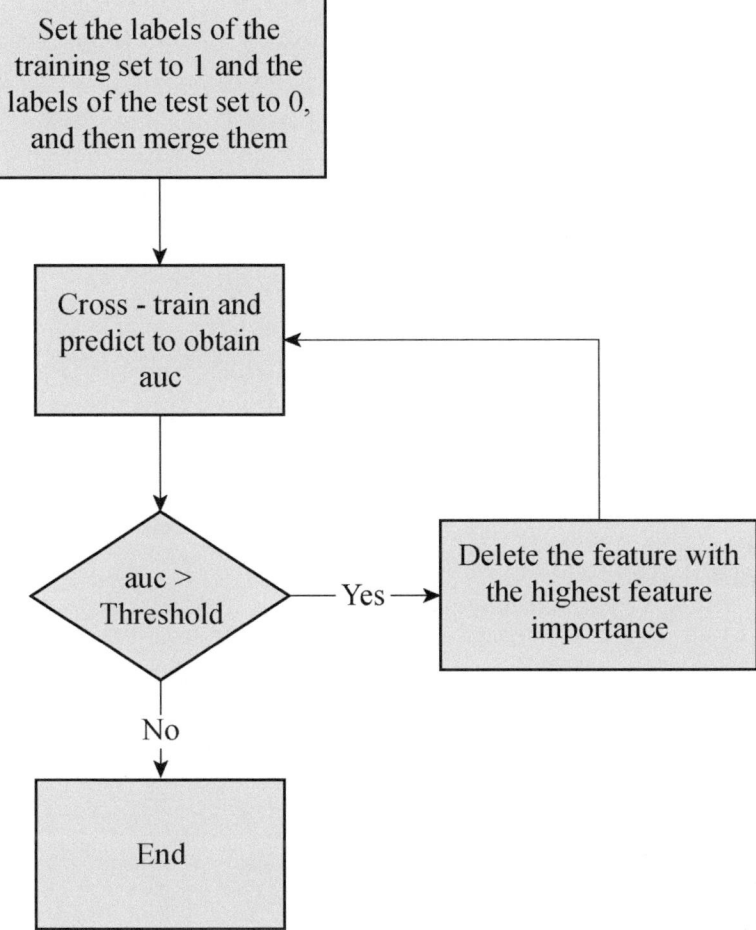

Fig. 2.13 Adversarial validation execution logic

```
        feature_importances = pd.DataFrame() # Create an empty
DataFrame to store feature importances
                feature_importances['feature']    =   train_
test[used].columns # Add the feature name column

            for  fold_n,  (train_index,  valid_index)  in
enumerate(folds.split(train_test[used])):      #      Start
cross-validation

    model = RandomForestClassifier() # Initialize the random
forest classifier
        model.fit(train_test[used].iloc[train_index], train_
test[target].iloc[train_index]) # Fit the model
```

```
        feature_importances['fold_{}'.format(fold_n + 1)]
= model.feature_importances_ # Obtain and store the feature
importances of each fold

        val = model.predict(train_test[used].iloc[valid_
index]) # Make predictions

                 auc_score  =  roc_auc_score(train_
test[target].iloc[valid_index], val) # Calculate the AUC
score
    AUCs.append(auc_score) # Store the AUC score

  mean_auc = np.mean(AUCs) # Calculate the average AUC score
     print(f'Mean AUC: {mean_auc}') # Print the average AUC
score

   # Calculate the average importance of each feature
   feature_importances['average'] = feature_importances[
      [x for x in feature_importances.columns if x !=
"feature"]].mean(axis=1)
   # Sort the features according to the average importance,
with features of higher importance ranked first, and reset the
index
        feature_importances = feature_importances.sort_
values(by="average",                ascending=False).reset_
index(drop=True)

   # If the average AUC score is greater than the set threshold,
then remove the most important feature and repeat the above
process
   if mean_auc > threshold:
        cur_removed_feature = feature_importances.loc[0,
'feature']
     print(f"remove feature {cur_removed_feature}")
     removed_features.append(cur_removed_feature)
     used = [x for x in used if x not in removed_features]
   else: # If the average AUC score is less than or equal to the
set threshold, then exit the loop and return the list of removed
features
     return removed_features
```

2.5   Model

In machine learning, a model refers to the rules or functions learned from data that can be utilized for prediction and classification. In this section, we will introduce some commonly used structured data models, including gradient boosting decision trees, neural network models, and others. Additionally, we will explore how to optimize model performance by adjusting hyperparameters and introduce some offline validation techniques to evaluate the accuracy and generalization ability of the model.

2.5.1 Commonly Used Structured Data Models

1. Gradient Boosting Decision Trees

Gradient boosting decision trees (GBDTs) are models constructed using decision trees as the base model, combined with the boosting ensemble strategy. They are currently the model first considered for structured data problems and can help users quickly obtain a well-performing baseline models. The advantages of GBDT are as follows.

- No need to scale features.
- Can automatically handle missing values.
- Can reflect variable interaction information without explicit cross-variables.
- Having unnecessary features does not significantly harm model performance.
- Can handle sparse matrix objects.

Below are three representative algorithm implementation tools or frameworks based on GBDT.

(1) XGBoost

XGBoost[1] is an open-source GBDT framework. Since 2015, XGBoost has been widely used and recognized in the data competition community, with a considerable number of top-ranking solutions using XGBoost.

XGBoost has made many improvements to the traditional GBDT, resulting in excellent performance in terms of effectiveness, efficiency, flexibility, and scalability.

- The objective function includes a regularization term to alleviate overfitting.
- The calculation process uses the second-order Taylor expansion, which allows the model to converge faster and facilitates custom loss functions, enabling the model to support a wide variety of loss functions.
- The sparse-aware algorithm reduces computation time and memory consumption when dealing with data containing a large number of missing values.
- Weighted approximate quantile sketch proposes an online, weighted quantile calculation method that can obtain an approximate solution with a theoretical error not exceeding a certain range, thereby greatly reducing computational complexity.
- It preemptively uses block structure (sparse matrix storage format CSC) to store the sorted values of all continuous features, which can be directly called to improve efficiency.
- Feature computation is performed using multi-threaded parallel computation.

---

[1] The GitHub URL of XGBoost is https://github.com/dmlc/XGBoost.

```
params = {
     'objective': 'reg:linear',
     'eval_metric': 'rmse',
     'learning_rate': 0.1,
     'max_depth': 2,
}

trn_data = xgb.DMatrix(X_train, label=y_train)
val_data = xgb.DMatrix(X_valid, label=y_valid)

clf = xgb.train(params, trn_data,

          num_boost_round=2000,
          evals=[(trn_data, 'train'), (val_data, 'valid')],
          verbose_eval=50,
          early_stopping_rounds=10)

xgb_preds = clf.predict(xgb.DMatrix(X_test))
xgb_rmse     =      mean_squared_error(y_test,      xgb_preds,
squared=False)
```

(2) LightGBM

LightGBM is a GBDT framework released by Microsoft, emphasizing lightness and achieving effects comparable to XGBoost with higher efficiency.[2]

LightGBM mainly improves algorithm efficiency through the following techniques.

- Gradient-Based One-Side Sampling (GOSS): Samples with large gradients are retained for training the next tree, while samples with smaller gradients are selected by sampling to determine whether to enter the next round of training samples.
- Leaf-Wise Growth Strategy: Unlike XGBoost's level-wise growth strategy, LightGBM uses a leaf-wise approach to select the split method of decision trees, which is a more efficient strategy that helps LightGBM fit more quickly.
- Exclusive Feature Bundling (EFB) involves bundling multiple mutually exclusive sparse features together into one feature to reduce dimensionality and improve efficiency.

```
params = {
     'objective': 'regression',
     'metric': 'rmse',
     'boosting': 'gbdt',
     'learning_rate': 0.1,
     'max_depth': 2
     }
```

---

[2] The GitHub URL of LightGBM is https://github.com/Microsoft/LightGBM.

```
trn_data = lgb.Dataset(X_train, label=y_train, categorical_
feature=
cat_features)
val_data = lgb.Dataset(X_valid, label=y_valid, categorical_
feature=
cat_features)

lgb1 = lgb.train(params, trn_data,
        num_boost_round=2000,
        valid_sets=[trn_data, val_data],
        verbose_eval=50,
        early_stopping_rounds=20)

lgb_preds = lgb1.predict(X_test)
lgb_rmse    =     mean_squared_error(y_test,    lgb_preds,
squared=False)
```

(3) CatBoost

CatBoost is another open-source GBDT framework released by the Russian company Yandex.[3] In CatBoost, "cat" stands for category, and its key feature is the ability to handle categorical features in data well, thereby improving performance.

Compared to XGBoost and LightGBM, CatBoost's innovations are as follows.

- Target encoding for categorical variables.
- Constructing combination features of categorical variables (also with target encoding).
- Using a completely symmetric tree as the base model.

```
train_pool = Pool(X_train, label=y_train, cat_features=cat_
features)
valid_pool = Pool(X_valid, label=y_valid, cat_features=cat_
features)

cat1 = CatBoostRegressor(iterations=1000,
            loss_function='RMSE',
            eval_metric='RMSE',
            metric_period=50,
            max_depth=3,
            early_stopping_rounds = 20
            )
cat1.fit(train_pool, eval_set=valid_pool)

cat_preds = cat1.predict(X_test)
cat_rmse    =     mean_squared_error(y_test,    cat_preds,
squared=False)
```

---

[3] The link is https://catboost.ai/.

2. Neural Network Models

Currently, in machine learning competitions with structured data, GBDT series models still dominate, and in most scenarios, neural network models struggle to compete with them. There may be two reasons for this situation: First, neural network models excel at representation learning, and their advantage lies in extracting features from unstructured data, whereas structured data has already been represented as features (such as petal length and width), and using neural network models on these features does not offer a significant advantage. Second, most structured competition datasets are relatively small, and neural network models have many parameters, requiring a large amount of data to fit well.

Nevertheless, in some competitions, neural network models can still perform well, sometimes achieving results close to or even surpassing those of GBDT series models. Additionally, neural network models can also serve as a very important base model in ensemble learning.

There are two main ways to build neural network models for structured data: One is to use open-source deep learning frameworks (such as TensorFlow, Keras, and PyTorch) to build them, where you can design the network structure yourself or refer to some classic network structures (such as Wide and Deep); the other is to use packaged neural network model libraries, such as TabNet and TabPFN. Here, TabNet is used as an example to demonstrate how to model structured data, with the code as follows.

```
clf = TabNetClassifier()
clf.fit(
  X_train, y_train,
  eval_set=[(X_valid, y_valid)]
)

tabnet_preds = clf.predict(X_test)
tabnet_auc = roc_auc_score(y_test, tabnet_preds)
```

3. Other Models

Machine learning has many other models. Compared to the GBDT series and neural network models, these models' fitting ability is relatively weaker, and they are usually not used alone. However, when ensemble learning is needed, they can be considered to increase model diversity (increasing model diversity is beneficial for ensemble learning). Common other models include linear models, support vector machines, random forests, etc., which can be implemented and called through Scikit-learn.[4]

4. Key Parameters of Common Models for Structured Data

The key parameters of common models for structured data are shown in Table 2.6.

---

[4] The link is https://scikit-learn.org/stable/index.html.

Table 2.6 Key parameters of common models for structured data

Model	Key parameters
Logistic regression	C: The inverse of regularization strength. Must be a positive float. The smaller the value, the stronger the regularization strength
Support vector machine (SVM)	C: Refer to the C in the logistic regression model kernel: Used to specify the kernel function of the algorithm. Includes types such as linear, poly, rbf, sigmoid, precomputed degree: When the kernel function is a polynomial kernel (poly), it is used to specify the degree of the polynomial. Generally set to 2–5. The larger the value, the stronger the model's fitting ability, and the easier it is to overfit gamma: When the kernel function is poly, rbf, or sigmoid, this parameter can be used. Gamma can be set to scale, auto, or a float. When set to scale or auto, the default method is used to calculate value; when set to a float, the larger the value, the stronger the model's fitting ability, and the easier it is to overfit
Random forest	n_estimators: The number of decision trees. Generally set to a positive integer between 10 and 2000. Since the random forest model is not easy to overfit, the larger the value, the better the model effect, but the data volume and computing resources need to be considered max_depth: The maximum tree depth of the decision tree. Generally set to a positive integer between 1 and 10. The larger the value, the more complex the model, the stronger the fitting ability, and the easier it is to overfit min_samples_split: The minimum number of samples required to split a node. Generally set to a positive integer between 1 and 300. The larger the value, the more it can prevent overfitting min_samples_leaf: The minimum number of samples required for a leaf node. Generally set to a positive integer between 0 and 300. The larger the value, the smoother the model, and the more it can prevent overfitting bootstrap: Whether to use the bootstrap method to construct decision trees. If False, the entire dataset is used to construct each decision tree

(continued)

Table 2.6 (continued)

Model	Key parameters
LightGBM	num_iterations: The number of iterations for boosting. Generally chosen between 10 and 10,000 based on the sample size and number of features. The larger the number of iterations, the stronger the fitting effect, and the easier it is to overfit. It is recommended to set a large number of iterations and use early stopping to let the model automatically choose the appropriate number of iterations
	learning_rate: The learning rate, generally set between 0.01 and 1.0. The larger the learning rate, the faster the model updates, but it is also more likely to diverge, affecting model convergence; the smaller the learning rate, the slower the model updates, but it is easier to achieve stable model performance
	feature_fraction: During each iteration of the model, a random portion of features is selected for fitting. This parameter indicates the proportion of selected features to the total features, limited to 0–1. A smaller proportion can speed up training and prevent overfitting, while a larger proportion results in stronger fitting ability
	num_leaves: The number of leaf nodes on a tree. Generally set to $2-2^{\text{max\_depth}}$. Since LightGBM uses a leaf-wise algorithm, it uses num_leaves instead of max_depth. The approximate conversion relationship is num_leaves $= 2^{\text{max\_depth}}$, but its setting should be slightly less than $2^{\text{max\_depth}}$. The larger the num_leaves, the stronger the fitting ability, but it is also easier to overfit
	subsample: During each iteration, a random portion of data is selected for fitting without resampling. This parameter indicates the proportion of selected data to the total samples, limited to 0–1. A smaller proportion can speed up training and prevent overfitting, while a larger proportion results in stronger fitting ability
	reg_alpha: L1 regularization coefficient generally set to a float between 0 and 100. The larger the value, the more it can prevent overfitting
	reg_lambda: L2 regularization coefficient, generally set as a floating-point number between 0 and 100. The larger the value, the better it prevents overfitting
	min_data_in_leaf: The minimum number of data in a leaf, generally set as an integer between 1 and 300. The larger the value, the better it prevents overfitting

(continued)

Table 2.6 (continued)

Model	Key parameters
XGBoost	n_estimators: Refer to LightGBM's num_iterations
	learning_rate: Refer to LightGBM's learning_rate
	min_child_weight: The minimum sum of instance weight (hessian) needed in a child. If the splitting process results in a leaf node with the sum of instance weight less than min_child_weight, then the building process will give up further partitioning. Generally set as an integer between 1 and 10. The larger the value, the better it prevents overfitting
	max_depth: Maximum tree depth, generally set as an integer between 1 and 10. The larger the value, the more complex the model, the stronger the fitting ability, but also more prone to overfitting
	subsample: Refer to LightGBM's subsample
	colsample_bytree: The proportion of features sampled when constructing each tree, limited between 0 and 1. A smaller proportion can speed up training and prevent overfitting, while a larger proportion results in stronger fitting ability
	colsample_bylevel: The proportion of features sampled when constructing a new level of the decision tree, limited between 0 and 1. A smaller proportion can speed up training and prevent overfitting, while a larger proportion results in stronger fitting ability
	reg_lambda: Refer to LightGBM's reg_lambda
	reg_alpha: Refer to LightGBM's reg_alpha
	gamma: The minimum loss reduction required to make a further partition on a leaf node, generally set between $1e-9$ and 0.5. The larger the value, the less likely the model is to overfit
	scale_pos_weight: Represents the weight of positive samples, used to control the weight of positive and negative samples. Useful when samples are imbalanced. Set according to the ratio of positive to negative samples in the data, a typical value to consider is the total number of negative examples divided by the total number of positive examples

(continued)

Table 2.6 (continued)

Model	Key parameters
CatBoost	iterations: The number of boosting iterations, settings can refer to LightGBM's num_iterations
	depth: The depth of the tree, generally set as an integer between 1 and 10. The larger the value, the more complex the model, the stronger the fitting ability, but also more prone to overfitting
	learning_rate: Refer to LightGBM's learning_rate
	random_strength: When the tree model splits, it scores each possible split (e.g., how much this split can reduce the loss function on the training set), then sorts all scores and selects the split with the highest score. This parameter represents the multiplication coefficient of the variance corresponding to this random variable. Generally set as a floating-point number between 1e−9 and 10.0. The larger the value, the better the effect of combating overfitting, but it also reduces the model's fitting ability
	border_count: The number of bins for continuous variables, limited between 1 and 65,535, generally set as an integer between 1 and 255. The larger the value, the stronger the fitting ability for continuous variables
	l2_leaf_reg: L2 regularization coefficient, generally set as an integer between 2 and 30. The larger the value, the better it prevents overfitting
	scale_pos_weight: Refer to XGBoost's scale_pos_weight

2.5.2   Model Hyperparameter Optimization

In the process of creating machine learning models, it is necessary to define some hyperparameters of the model architecture to determine the model's structure. Usually, we cannot immediately know the best model architecture, so we need to explore these hyperparameters in the hope of obtaining a model architecture with high accuracy and good generalization performance. Below are some commonly used hyperparameter optimization search methods for tabular data.

1. Grid Search

Grid search defines the search space as a grid of hyperparameter values and then traverses and evaluates each point in the grid.

```
parameters = {
    'max_depth': [2,3,4,5,6],
    'min_samples_split': [2,3],
    'min_samples_leaf': [2,3],
    'min_weight_fraction_leaf': [0, 0.1, 0.2]
}

clf = GridSearchCV(
    RandomForestClassifier(random_state=42),
    parameters, refit=True, verbose=1,
)
clf.fit(x_train, y_train)

# Print the optimal parameters
print(clf.best_params_)

# Evaluate the test set using the optimal parameters
print(clf.best_estimator_.score(x_test, y_test))
```

2   Random Search

Random search defines the search space as a bounded domain of hyperparameter values and randomly samples within this domain for searching. Random search can explore the hyperparameter space more extensively. As shown in Fig. 2.14, the hyperparameter on the horizontal axis is more important than that on the vertical axis. With grid search, only 3 different hyperparameter values can be searched on the horizontal axis, while the random search method explores 9 different values on the horizontal axis.

```
parameters = {
    'max_depth': [2, 3, 4, 5, 6],
    'min_samples_split': [2, 3],
    'min_samples_leaf': [2, 3],
```

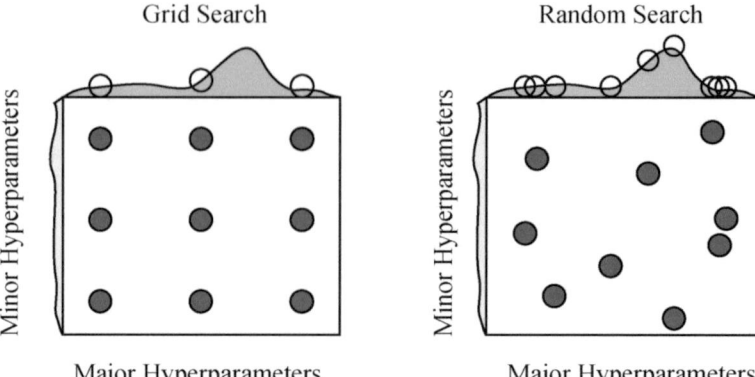

Fig. 2.14 Grid search and random search

```
    'min_weight_fraction_leaf': uniform(loc=0.1, scale=0.3)
}

clf = RandomizedSearchCV(
    RandomForestClassifier(random_state=42),
    parameters, refit=True, verbose=1, n_iter=10
)

clf.fit(x_train, y_train)

# Print the optimal parameters
print(clf.best_params_)

# Evaluate the test set using the optimal parameters
print(clf.best_estimator_.score(x_test, y_test))
```

3. Halving Search

Halving search uses the successive halving strategy to search for hyperparameters. At the beginning of the iteration, a small amount of data is used to evaluate all candidate hyperparameter combinations. In the next iteration, only the better-performing half of all candidate hyperparameter combinations are selected, and they will be evaluated with more data. As each iteration progresses, the excellent candidate parameter combinations will receive more and more data. The halving search strategy can be combined with grid search or random search.

(1) Halving Grid Search

Halving grid search is a combination of halving search and grid search. The example code is as follows:

```
parameters = {
    'max_depth': [2, 3, 4, 5, 6],
  'min_samples_split': [2, 3],
  'min_samples_leaf': [2, 3],
  'min_weight_fraction_leaf': [0, 0.1, 0.2]
}

clf = HalvingGridSearchCV(
    RandomForestClassifier(random_state=42),
    parameters, refit=True, verbose=1
)
clf.fit(x_train, y_train)

# Print the optimal parameters
print(clf.best_params_)

# Evaluate the test set using the optimal parameters
print(clf.best_estimator_.score(x_test, y_test))
```

(2) Halving Random Search

Halving random search is a combination of halving search and random search. The example code is as follows:

```
parameters = {
    'max_depth': [2, 3, 4, 5, 6],
  'min_samples_split': [2, 3],
  'min_samples_leaf': [2, 3],
  'min_weight_fraction_leaf': uniform(loc=0.1, scale=0.3)
}

clf = HalvingRandomSearchCV(
    RandomForestClassifier(random_state=42),
    parameters, refit=True, verbose=1
)

clf.fit(x_train, y_train)

# Print the optimal parameters
print(clf.best_params_)

# Evaluate the test set using the optimal parameters
print(clf.best_estimator_.score(x_test, y_test))
```

4. Bayesian Optimization

The key idea behind Bayesian optimization is to optimize a proxy function rather than the true objective function. The steps to create a typical Bayesian optimization pseudo-code are as follows:

(1) Define the objective function $y = f(x)$. Here, x represents the hyperparameters of the model, and y represents the score of the corresponding hyperparameters on the validation set.

(2) Obtain the cold start data $[X, Y]$. Suppose there are 20 pieces of data, i.e., $(x_1, y_1), (x_2, y_2), \ldots, (x_{20}, y_{20})$.

(3) Fit the data using a Gaussian mixture model, GP.fit(X, Y).

(4) Randomly generate 100 sets of parameters: $\tilde{x}_1, \tilde{x}_2, \ldots, \tilde{x}_{100}$. Use the Gaussian model to make predictions on these 100 sets of parameters, and obtain the predicted values and standard deviations, which correspond to $\tilde{y}_1, \tilde{y}_2, \ldots, \tilde{y}_{100}$ and $\widetilde{std}_1, \widetilde{std}_2, \ldots, \widetilde{std}_{100}$.

(5) The score p corresponding to the generated 100 sets of random parameters is $\tilde{p}_i = \frac{\tilde{y}_i - y_{\text{best}}}{\widetilde{std}_i + \varepsilon}$. For i from 1 to 100, obtain the parameter $\tilde{x}_i$ corresponding to the optimal p_i. Here, ε represents a very small number, and y_{best} represents the best score: $y_{\text{best}} = \max(y_1, y_1, \ldots, y_{20})$ corresponding to the currently known optimal parameters.

(6) Train with the parameter $\tilde{x}_i$ and then obtain the corresponding score on the validation set.

(7) Update $[X, Y]$ with the newly obtained samples from step (6), then go back to step (3) to retrain the model, and iterate the operations from step (3) to step (7).

2.5.3   Offline Validation

The purpose of constructing an offline validation set in machine learning competitions is as follows.

(1) Improve efficiency. Most machine learning competitions limit the number of submissions, and during the modeling process, it is necessary to frequently modify algorithm configuration parameters, etc. If feedback is obtained solely through online submissions, the efficiency is very low.

(2) Guide the direction of modeling through a reasonable validation set, such as data preprocessing, feature engineering, and model parameters.

The guiding principle for constructing a reasonable offline validation set is to make the relationship between the training set and the test set consistent with the relationship between the offline training set and the offline validation set. Depending on whether there is a temporal relationship between the training set and the test set, the offline validation set can be divided into non-sequential and sequential types.

1. Non-sequential

(1) KFold

KFold is a commonly used offline validation method. As shown in Fig. 2.15, it randomly divides the training set into approximately equal k parts, using $(k - 1)$

parts for training each time, and the remaining 1 part for validation, iterating k times, and taking the average of the k validation results as the offline validation result.

```
sub = test[['id']]
sub[target] = 0
AUCs = []

n_fold = 5
folds = KFold(n_splits = n_fold)

for train_index, valid_index in folds.split(train[used_
features]):

    trn_x, trn_y = train[used_features].iloc[train_index],
train[target].iloc[train_index]
    val_x, val_y = train[used_features].iloc[valid_index],
train[target].iloc[valid_index]

    model = LogisticRegression()
    model.fit(trn_x, trn_y)

    val_pred = model.predict(val_x)

    pred = model.predict(test[used_features])
    sub[target] = sub[target] + pred / n_fold

    auc_score = roc_auc_score(val_y, val_pred)
    AUCs.append(auc_score)

print(f'mean auc: {np.mean(AUCs)}')
```

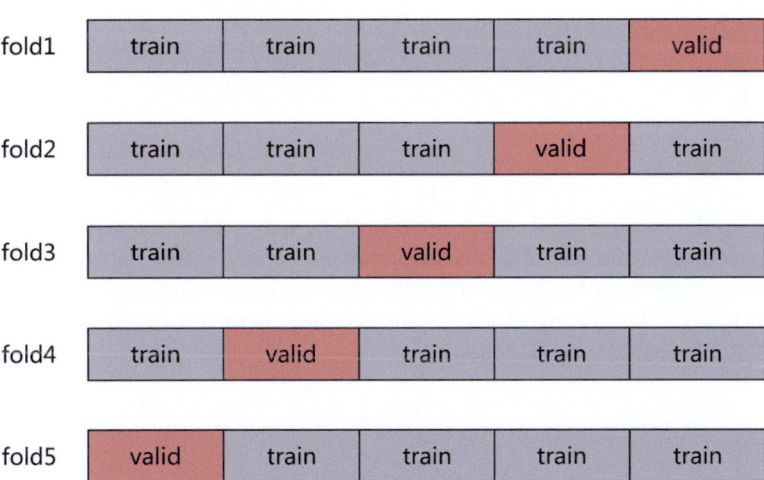

Fig. 2.15 Constructing an offline validation set with KFold cross-validation

(2)   StratifiedKFold

StratifiedKFold, building on KFold, performs stratified sampling, controlling the proportion of each category of data in each group to be relatively consistent. Therefore, in scenarios where the distribution of label categories is imbalanced, using StratifiedKFold may achieve better offline validation results than KFold.

```python
sub = test[['id']]
sub[target] = 0
AUCs = []

n_fold = 5
skf = StratifiedKFold(n_splits = n_fold)

for   train_index,   valid_index   in   skf.split(train[used_
features],
train[target]):

    trn_x, trn_y = train[used_features].iloc[train_index],
train[target].iloc[train_index]
    val_x, val_y = train[used_features].iloc[valid_index],
train[target].iloc[valid_index]

  model = LogisticRegression()
  model.fit(trn_x, trn_y)

  val_pred = model.predict(val_x)

  pred = model.predict(test[used_features])
  sub[target] = sub[target] + pred / n_fold

  auc_score = roc_auc_score(val_y, val_pred)
  AUCs.append(auc_score)
print(f'mean auc: {np.mean(AUCs)}')
```

(3)   GroupKFold

GroupKFold, building on KFold, can control the data within the same group to be divided into the same fold. For instance, when working with a dataset where each user has multiple samples, and you need to perform cross-validation by grouping all samples of a single user together (to avoid splitting them across different folds), the GroupKFold method is applicable.

```python
sub = test[['id']]
sub[target] = 0
AUCs = []

n_fold = 5
group_kfold = GroupKFold(n_splits = n_fold)
```

```
for        train_index,        valid_index        in        group_
kfold.split(train[used_features],            train[target],
train['user_id']):

    trn_x, trn_y = train[used_features].iloc[train_index],
train[target].iloc[train_index]
    val_x, val_y = train[used_features].iloc[valid_index],
train[target].iloc[valid_index]

    model = LogisticRegression()
    model.fit(trn_x, trn_y)

    val_pred = model.predict(val_x)

    pred = model.predict(test[used_features])
    sub[target] = sub[target] + pred / n_fold

    auc_score = roc_auc_score(val_y, val_pred)
    AUCs.append(auc_score)

print(f'mean auc: {np.mean(AUCs)}')
```

2. Time Series

A time series scenario refers to when the data collection time of the training set is earlier than that of the test set. For example, if the training set consists of daily sales data from January 2022 to November 2023, and you need to predict the daily sales in December 2023 in the test set, this is a typical time series scenario.

In a time series scenario, to ensure the model's generalization ability on the test set, reasonable offline validation is necessary. Below are two offline validation strategies for time series data.

(1) Time Series Data Offline Validation Strategy One

The first strategy is to select the last period of the training set as the offline validation set. This is a relatively easy way to split the offline validation set. Since the last period of the training set is the closest data to the test set, in most cases, it is also more similar to the test set. In the above sales prediction example, you can use the data from November 2023 as the offline validation set, as shown in Fig. 2.16.

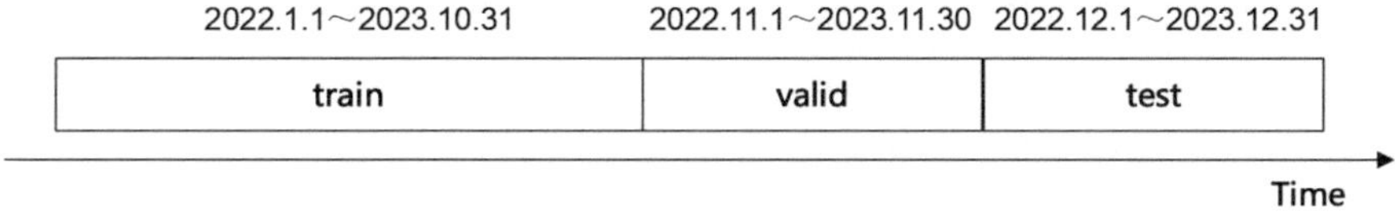

Fig. 2.16 First method of constructing a time series data validation set

```
local_train = train.loc[(train['date'] >= '2022-01-01') &
(train['date'] <= '2023-10-31')]
local_valid = train.loc[(train['date'] >= '2023-11-01') &
(train['date'] <= '2023-11-30')]
```

(2)   Time Series Data Offline Validation Strategy Two

The second method for constructing an offline validation set for time series data is to select a period that is relatively consistent with the distribution of the test set as the offline validation set. When the data is periodic, selecting the last period of the training set as the offline validation set may result in the offline validation set and the test set having inconsistent distributions, making the offline validation set less responsentative. In this case, you can choose a period that is more consistent with the test set distribution as the offline validation set. For example, in the aforementioned sales forecasting case, if the data has a similar trend on a yearly cycle, you can try using December of the previous year as the offline validation set. In this case, the offline training set would be the data from January 2022 to November 2022, and the offline validation set would be the data from December 2022, as shown in Fig. 2.17.

```
local_train = train.loc[(train['date'] >= '2022-01-01') &
(train['date'] <= '2022-11-31')]
local_valid = train.loc[(train['date'] >= '2022-12-01') &
.(train['date'] <= '2022-12-31')]
```

3.   Determine Whether the Constructed Offline Validation Set Is Reasonable

You can determine whether the constructed offline validation set is reasonable by evaluating the metrics on the offline validation set and the test set. If the difference between the two does not vary significantly across multiple results, then the construction of the offline validation set can be considered relatively reasonable. If there is some variation in the difference, but the trend (the relative order of the evaluation metrics across multiple results) is maintained, then the construction of the offline validation set can be considered basically reasonable.

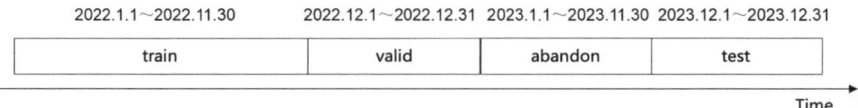

Fig. 2.17 Second method for constructing a time series data validation set

2.6 Ensemble Learning

Ensemble learning combines multiple individual models to obtain the final predictive model. Ensemble learning usually achieves better results than individual models, and in data mining competitions, top-ranked solutions almost always use ensemble learning techniques.

2.6.1 Voting Method

The voting method refers to selecting the category with the most votes as the final prediction result based on the principle of the majority rule. The voting method is suitable for classification problems.

```
df['vote'] = df[['model_1', 'model_2', 'model_3']].mode(axis
= 1)
```

2.6.2 Average Method

The average method refers to taking the average of the prediction results of multiple individual models, suitable for regression problems. The average method includes the most basic arithmetic mean and some variants (including geometric mean, harmonic mean, log-transformed mean, and n-power mean).

1. Arithmetic Mean

The arithmetic mean is the sum of the prediction results of multiple individual models divided by the number of individual models, as shown in the code below.

```
df['arithmetic_mean'] = df[['model_1', 'model_2', 'model_
3']].mean(axis = 1)
```

2. Geometric Mean

The geometric mean is the nth root of the product of the prediction results of n individual models, as shown in the code below.

```
df['geometric_mean'] = stats.gmean(df[['model_1', 'model_
2', 'model_3']], axis=1)
```

3. Harmonic Mean

The harmonic mean is calculated by taking the reciprocal of the arithmetic mean of the reciprocals of all individual model prediction results, as shown in the code below.

```
df['harmonic_mean'] = stats.hmean(df[['model_1', 'model_2',
'model_3']], axis=1)
```

4. Log-Transformed Mean

The log-transformed mean refers to taking the arithmetic mean of the log-transformed prediction results of all individual models and then taking the exponential to obtain the log-transformed mean, as shown in the code below.

```
def log_mean(preds):
return np.exp(np.mean([np.log(pred) for pred in preds]))
df['log_mean']   =   df[['model_1',   'model_2',   'model_
3']].apply(lambda x: log_mean(x), axis=1)
```

5. n-th power Mean

The n-th power mean refers to calculating the arithmetic mean of the n-th powers of all individual model predictions, and then taking the n-th root to obtain the n-power mean, as shown in the code below.

```
def npower_mean(preds, n):
return  np.power(np.mean([np.power(pred,  n)  for  pred  in
preds]), 1/n)
df['3power_mean']   =   df[['model_1',   'model_2',   'model_
3']].apply(lambda x: npower_mean(x, 3), axis=1)
```

2.6.3   Weighted Average Method

The weighted average method does not treat all individual models equally, but in practice, different models are usually assigned different weights. When using the weighted average method, it is necessary to consider how to assign reasonable weights to different models. Common methods include assigning weights based on model performance, correlation, and offline validation weight adjustment. It should be noted that the weighted average method can be combined with any of the aforementioned averaging methods; for illustration purposes, the arithmetic mean is used as an example.

1. Assigning Weights Based on Ranking

An intuitive idea is to assign different weights based on the scores of each individual model on the leaderboard, with higher-ranked models receiving higher weights. For example, a decreasing arithmetic sequence can be constructed based on model scores, then normalized to serve as the weights for the models. The example code is as follows.

```
rank = ['model_2', 'model_3', 'model_1']
w = np.array(range(3, 0, -1))
w = w / sum(w)
df['rank_weighted'] = df[rank].dot(w)
```

2. Assigning Weights Based on Correlation

The correlation of prediction results from different models varies. High correlation indicates that the prediction results of two models are similar. To more effectively comprehensively consider various models and increase the diversity of the ensemble model, models that are less correlated should be given higher weights. The idea of assigning weights based on correlation is as follows.

(1) Calculate the similarity matrix using the prediction results of individual models.
(2) Set the diagonal elements of the similarity matrix to 0.
(3) Convert the result of (2) to a vector by taking the mean of each row.
(4) Take the reciprocal of the result from (3).
(5) Normalize the result from (4) so that the sum is 1, and the resulting values are the weight of each individual model.

```
def corr_weight(df):
    corr_matrix = np.array(df.corr()) # Calculate the correla-
tion matrix of the input DataFrame
    np.fill_diagonal(corr_matrix, 0.0) # Set the elements on the
diagonal to 0 (since they are the correlations of each variable
with itself, which have a value of 1)
    w = np.mean(corr_matrix, axis=1) # Calculate the mean of each
row, that is, the average correlation of each feature with other
features
    w = 1 / w # Take the reciprocal because we expect features with
low correlations with other features to have higher weights
```

```
    w = w / np.sum(w) # Normalize the weights so that the sum of
all weights is 1
    return df.dot(w) # Calculate the weighted average of each
feature according to the weights

df['corr_weighted']    =    corr_weight(df[['model_1','model_
2','model_3']]) # Calculate the 'corr_weighted' feature
```

3. Weight Adjustment Based on Offline Validation

Another reasonable method is to adjust the weights of each individual model based on
the evaluation metrics of the offline validation set, which can help mitigate overfitting
of the ensemble model. The example code for obtaining the optimal weights using
fivefold cross-validation is as follows.

```
kf   =   KFold(n_splits=5,   shuffle=True,   random_state=0)   #
Initialize 5-fold cross-validation
best_p1, best_p2, best_p3 = None, None, None # Initialize the
best parameter values
best_auc = 0.5 # Initialize the best AUC score

# Use a nested loop to iterate over all possible values of p1 and
p2, with values in [0, 1] and a step size of 0.1
for p1 in range(0, 11):
    p1 = p1 / 10 # Convert p1 to the range [0, 1]
    for p2 in range(0, 11):
        p2 = p2 / 10 # Convert p2 to the range [0, 1]
        p3 = 1 - p1 - p2 # Calculate the value of p3 such that p1 +
p2 + p3 = 1
        if p3 < 0: # If p3 is less than 0, break out of the inner loop
            break
        AUCs = [] # Create an empty list to store the AUC scores for
each fold
        # Iterate over each fold of the data
        for train_index, valid_index in kf.split(X):
            # Fit the three models
            model_1.fit(X[train_index, :], y[train_index])
            model_2.fit(X[train_index, :], y[train_index])
            model_3.fit(X[train_index, :], y[train_index])
            # Calculate the predicted values based on the predicted
probabilities of the models and the parameters p1, p2, p3
            y_pred = model_1.predict_proba(X[valid_index, :])[:,
1] * p1 + model_2.predict_proba(X[valid_index, :])[:, 1] * p2
+ model_3.predict_proba(X[valid_index, :])[:, 1] * p3
            # Calculate the AUC score
            auc_ = roc_auc_score(y[valid_index], y_pred)
            AUCs.append(auc_) # Store the AUC score
```

```
        # If the current average AUC score is greater than the
    best AUC score, then update the best parameters and the best AUC
    score
        if np.mean(AUCs) > best_auc:
            best_p1, best_p2, best_p3 = p1, p2, p3
            best_auc = np.mean(AUCs)
```

2.6.4　Stacking

Stacking is a very powerful model ensemble technique, and its overall architecture is shown in Fig. 2.18. Stacking consists of two layers of models. The first layer models are called base models, and their role is to generate meta-features. The input to the base models is the complete training set, and their output is the meta-features. The second layer model is called the meta-model, and its role is to generate the final prediction results. The input to the meta-model is the meta-features generated by the first layer, and its output is the final prediction results.

The base models generate meta-features through cross-validation, as shown in Fig. 2.19. Suppose there is a base model named model_1, and we use fivefold cross-validation as an example. Each time, four folds are used for training, and predictions are made on the remaining fold and the test set. This process is repeated five times to obtain all the prediction results on the training set, which are concatenated to form the meta-features of the base model on the training set. The meta-features of the test set are the average of the five prediction results. By constructing N base models, we obtain N sets of meta-features. Generally, it is hoped that the diversity of the base models is as large as possible. When the base models are significantly different, the

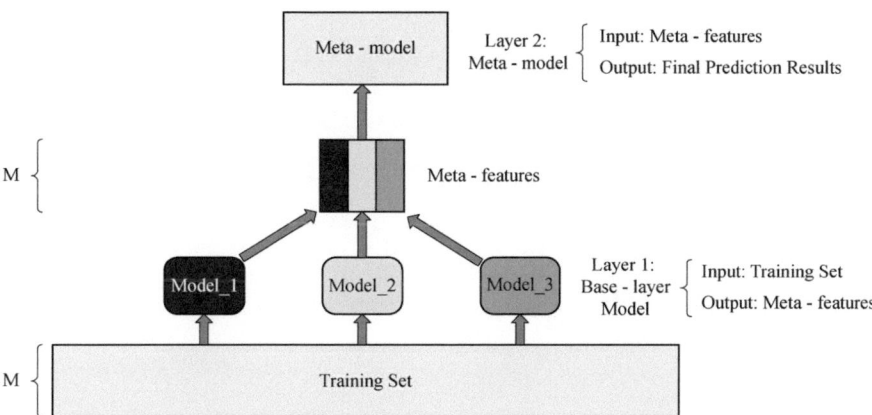

Fig. 2.18 Overall architecture of Stacking

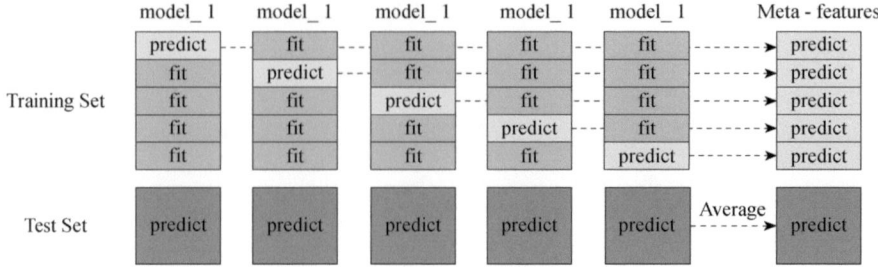

Fig. 2.19 Base model generating meta-features in Stacking

final effect of Stacking is better. Methods to construct diversity among base models include using different types of models (including linear models, tree models, neural network models, etc.), varying model hyperparameters, selecting different features, etc.

The meta-model is further trained based on the meta-features obtained from the base models and the true labels of the corresponding training set, and it predicts on the basis of the meta-features of the test set to obtain the final prediction results. Usually, the meta-model employs relatively simple models, such as linear models.

The code example for Stacking is as follows.

```
# Initialize three base models
model_1 = RandomForestClassifier(n_estimators=10, random_
state=42) # Random Forest Classifier
model_2 = LinearSVC(random_state=42) # Linear Support Vector
Machine Classifier
model_3 = GradientBoostingClassifier(random_state=42) #
Gradient Boosting Classifier

# Combine the base models into a list
estimators = [('model_1', model_1), ('model_2', model_2),
('model_3', model_3)]

# Initialize the stacking classifier
stacking_clf = StackingClassifier(
    estimators=estimators, # Base models
    final_estimator=LogisticRegression() # Meta-model
)

# Fit the stacking classifier using the training data
stacking_clf.fit(X_train, y_train)
# Make predictions on the test data
stacking_pred = stacking_clf.predict(X_test)
```

2.6.5 *Blending*

Consistent with the architecture of Stacking shown in Fig. 2.18, the overall architecture of Blending is also a two-layered structure, as shown in Fig. 2.20.

The first layer is the meta-feature construction module. Blending does not use the cross-training method in Stacking but instead splits the original training set into two parts. One part is used to train the base models, and the other part and the test set are used for prediction. The obtained prediction results serve as the meta-features of the first layer. Using multiple base models can generate multiple meta-features.

The second layer is the meta-learner module. It uses the meta-features of the first layer and the true labels of the corresponding data to train and to predict the final results based on the meta-features of the test set.

The code example for Blending is as follows.

```
# Prepare the data
X, y = make_classification(n_samples=2000) # Create a dataset
for classification
X_train_valid, X_test, y_train_valid, y_test = train_test_
split(X, y, test_size=0.2, random_state=42) # Split the data
into training + validation set and test set
X_train, X_valid, y_train, y_valid = train_test_split(X_
train_valid, y_train_valid, test_size=0.5, random_state=42)
# Further split the training + validation set into training set
and validation set

# Define the base models
```

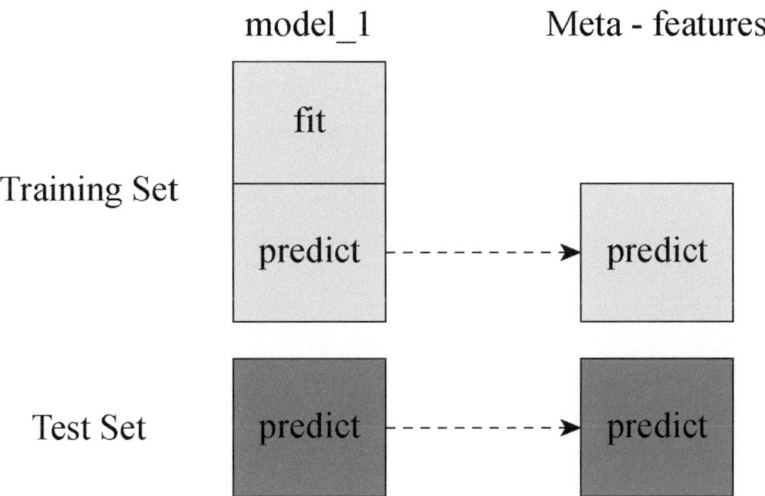

Fig. 2.20 Base model generating meta-features in Blending

```python
model_1 = RandomForestClassifier(n_estimators=10, random_
state=42) # Random Forest Classifier
model_2 = LinearSVC(random_state=42) # Linear Support Vector
Machine Classifier
model_3 = GradientBoostingClassifier(random_state=42) #
Gradient Boosting Classifier

# The first layer of Blending
models = [model_1, model_2, model_3] # Combine the base models
into a list

meta_train = np.zeros((len(X_valid), len(models))) # Create
an empty matrix to store the prediction results of the base
models on the validation set
meta_test = np.zeros((len(X_test), len(models))) # Create an
empty matrix to store the prediction results of the base models
on the test set

for i, model in enumerate(models): # For each base model
    model.fit(X_train, y_train) # Train the model using the
training set
    meta_train[:, i] = model.predict(X_valid) # Store the
model's prediction results on the validation set
    meta_test[:, i] = model.predict(X_test) # Store the model's
prediction results on the test set

# The second layer of Blending
meta_learner = LogisticRegression() # Define the meta-model,
using Logistic Regression here
meta_learner.fit(meta_train, y_valid) # Train the meta-model
using the prediction results of the base models on the valida-
tion set as input and the true results of the validation set as
output
blending_pred = meta_learner.predict(meta_test) # Use the
meta-model to make predictions on the test set
```

Chapter 3
Structured Data: Practical Part

This chapter uses the Home Credit Default Risk competition (see Fig. 3.1, image source is the competition homepage) as an example to introduce practical solutions for structured data competitions.[1]

3.1 Competition Overview

Many people find it difficult to obtain loans due to insufficient or non-existent credit records. Unfortunately, these individuals are often exploited by untrustworthy lenders. Home Credit aims to expand lending inclusivity. To verify whether the legitimacy and reasonableness of these individuals' loans, Home Credit hopes to use various data (including telecommunications and transaction information) to predict customers' repayment ability. Specifically, participants need to integrate various information to predict whether each loan will be repaid late, which is a typical binary classification problem.

The competition data includes a total of 8 tables. Each sample in the main table represents a loan application, where the main table of the training set includes a label column, and the main table of the test set does not. In addition to the main table, there are also records of customers' loans and repayments on other financial platforms, as well as their historical loans and repayments with Home Credit. The introduction of each table and the relationships between the tables are shown in Fig. 3.2.[2]

The evaluation metric for this competition is AUC (area under the ROC curve).

---

[1] The competition website is https://www.kaggle.com/competitions/home-credit-default-risk.

[2] The source of the picture is the data details page of the competition question, the link is https://www.kaggle.com/competitions/home-credit-default-risk/data.

Fig. 3.1 Home Credit Default Risk competition

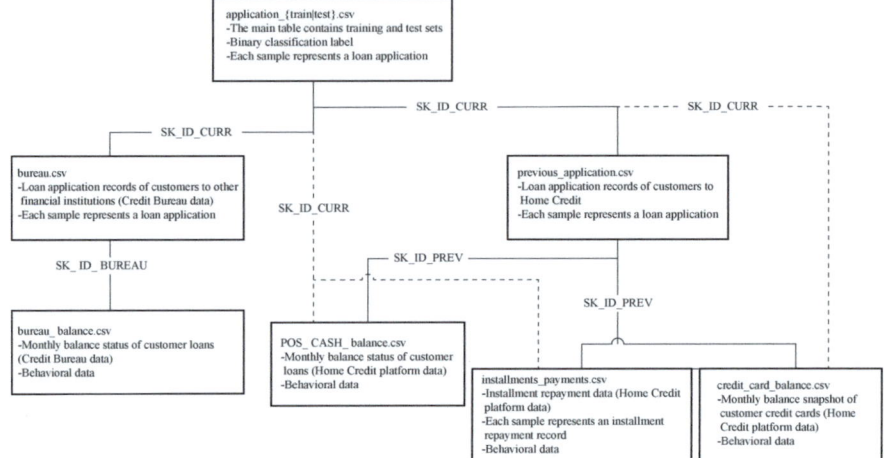

Fig. 3.2 Home Credit competition data relationship table

3.2 Data Exploration

Data exploration is an open-ended process that can help us discover trends, anomalies, patterns, and relationships in the data. These findings can guide the subsequent modeling process, including how to perform data preprocessing, which features to construct, and which models to choose. This section examines the competition data in four aspects: label distribution, missing values, outliers, and correlation.[3]

3.2.1 Label Distribution

This is a binary classification problem, where a label of 0 indicates the loan was repaid on time, and a label of 1 indicates the loan was not repaid on time. Figure 3.3 shows the distribution of labels in the training set, where the number of samples

---

[3] The reference link is https://www.kaggle.com/code/willkoehrsen/start-here-a-gentle-introduct ion#Exploratory-Data-Analysis.

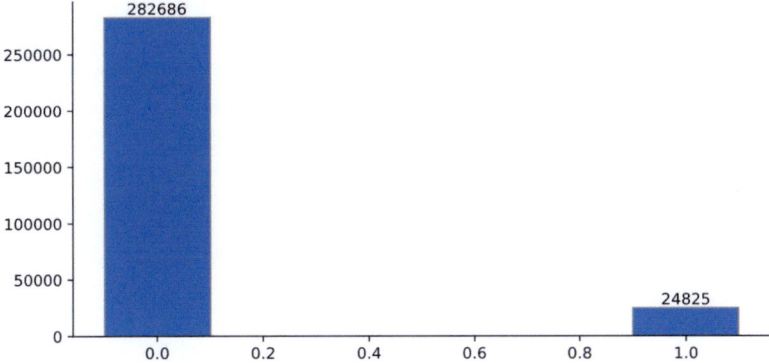

Fig. 3.3 Distribution of labels in the training set

repaid on time is 282,686, and the number of samples not repaid on time is 24,825, with a ratio exceeding 10:1. This indicates that this is a class-imbalanced dataset.

3.2.2 Missing Values

Analyze the missing values in the main table of the training set using the following code.

```
def missing_values_table(df):
    mis_val = df.isnull().sum()  # Calculate the number of
missing values in each column of the DataFrame
    mis_val_percent = 100 * df.isnull().sum() / len(df) # Calcu-
late the percentage of missing values in each column of the
DataFrame
    mis_val_table = pd.concat([mis_val, mis_val_percent],
axis=1)  # Combine these two Series into a new DataFrame
    mis_val_table_ren_columns = mis_val_table.rename(
        columns={0: 'Number of Missing Values', 1: 'Percent of
Total Values'})  # Rename the columns of the DataFrame
    # Keep only the columns with missing values and sort them in
descending order by the percentage of missing values
    mis_val_table_ren_columns = mis_val_table_ren_columns[
        mis_val_table_ren_columns.iloc[:, 1] != 0].sort_values(
        'Percent of Total Values', ascending=False).round(1)
    print("Your selected dataframe has " + str(df.shape[1]) + "
columns.\n"
        "There are " + str(mis_val_table_ren_columns.shape[0])
+
```

```
        " columns that have missing values.")  # Print summary
    information: the total number of columns in the DataFrame and
    the number of columns with missing values

        return mis_val_table_ren_columns  # Return the resulting
    DataFrame

    missing_values = missing_values_table(train) # Run this func-
    tion on the training data and save the result
    missing_values.head(5)   # Print the first five rows of the
    resulting DataFrame
```

The main table of the training set has a total of 122 columns, of which 67 columns contain missing values, with the highest percentage of missing values reaching 69.9%. Since this solution uses LightGBM and XGBoost models, there is no need to fill in the missing values for this stage.

3.2.3 Outliers

During data exploration, it is crucial to always pay attention to the presence of outliers in the data. Outliers may result from data entry errors, measurement errors, or other reasons. Below is a case of an outlier in the data: examine the DAYS_ EMPLOYED column in the main table, which indicates the current employment duration of the customer (in days, calculated by subtracting the loan application date from the employment start date, resulting in a negative number). Figure 3.4 shows the data distribution of the DAYS_EMPLOYED column.

```
    train['DAYS_EMPLOYED'].describe()
```

```
train['DAYS_EMPLOYED'].describe()

            count      307511.000000
            mean        63815.045904
            std        141275.766519
            min        -17912.000000
            25%         -2760.000000
            50%         -1213.000000
            75%          -289.000000
            max        365243.000000
            Name: DAYS_EMPLOYED, dtype: float64
```

Fig. 3.4 Data distribution of the DAYS_EMPLOYED column

From Fig. 3.4, it can be seen that the maximum value is 365,243 (approximately 1000 years), which is clearly illogical. Upon analysis, it was found that all samples with values greater than 0 in this column are 365,243, and there are no missing values in this column because this value—365,243—was assigned to customers without employment. We replace these outliers with NaN.

3.2.4 Correlation

The following statement uses the corr() method to calculate the Pearson correlation coefficient between each variable in the main table and the target.

```
correlations = train.corr()['TARGET']
```

A heatmap of the top five features with the highest absolute correlation values with the label is plotted, as shown in Fig. 3.5. The three features with the highest correlation are EXT_SOURCE_3, EXT_SOURCE_2, and EXT_SOURCE_1, which represent three credit scores from external data sources and will also be used in the construction of subsequent meta-features.

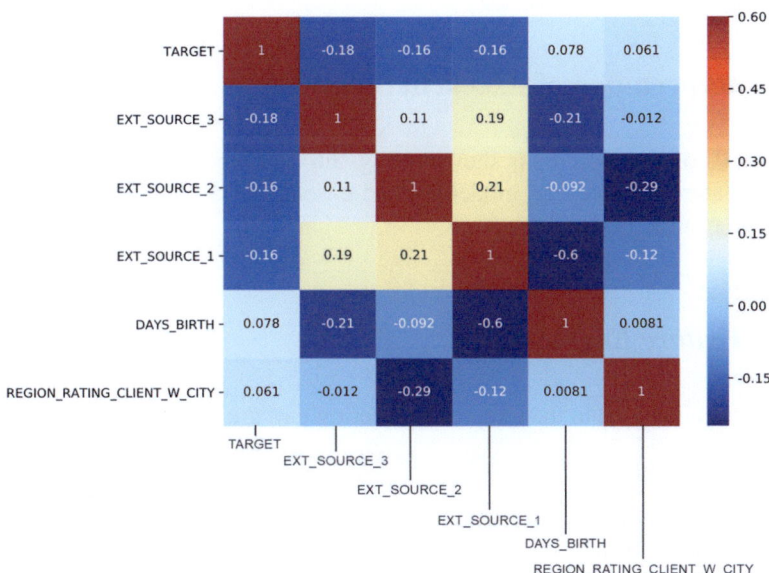

Fig. 3.5 Correlation heatmap

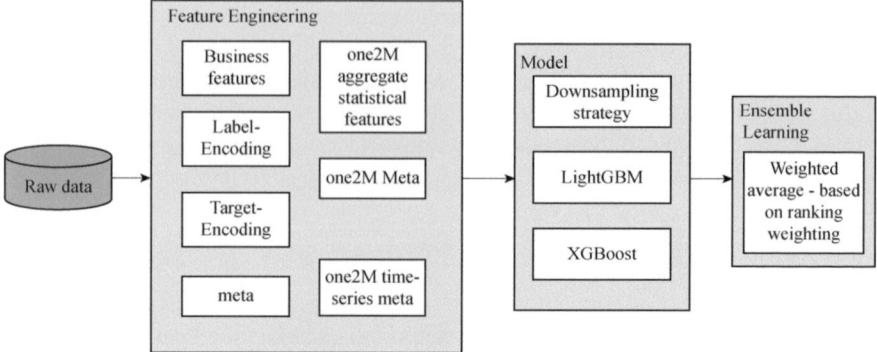

Fig. 3.6 Home Credit competition solution process flowchart

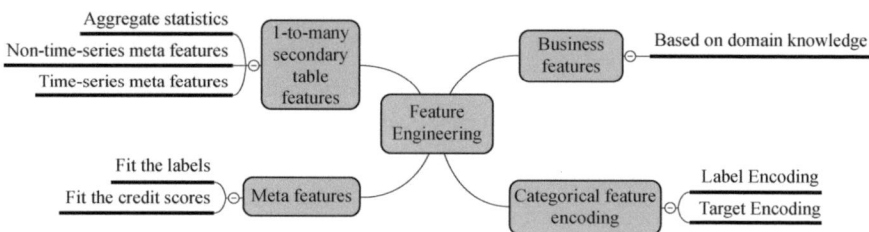

Fig. 3.7 Feature engineering modules

3.3   Interpretation of Excellent Solutions

The overall solution process is shown in Fig. 3.6, starting from the raw data, going through feature engineering, modeling, and ensemble learning, to obtain the final prediction results.[4]

3.3.1   Feature Engineering

The modules included in the feature engineering part are shown in Fig. 3.7.

1.   Business Features

Construct meaningful business features based on the understanding of domain knowledge, such as the difference between the monthly payable amount and the actual

---

[4] The reference link for this solution is https://github.com/NoxMoon/home-credit-default-risk. The reference link for the solution code introduced in this chapter is https://github.com/poteman/kaggle-home-credit-default-risk.

repayment amount, the time since the customer's last overdue payment, and the ratio of card spending to credit card limit.

2. Categorical Feature Encoding

The data contains many categorical features. For features with fewer categories, use Label Encoding to convert categorical data into numerical data. For features with more categories, use Target Encoding to convert categorical variables into the average value of the target under that category (statistics from the training set).

3. Meta-features

Here, meta-features refer to features that use the model's predicted values as features. To avoid overfitting of meta-features, a fivefold cross-validation method is used. The construction method of meta-features can be referred to in Sect. 2.6.4. In this case, the following three schemes are used to generate meta-features.

(1) Use features related to the customer's house in the main table and fit the label using a linear regression model.
(2) Use features related to the customer's house in the main table and fit the user's credit score (calculated as the average of EXT_SOURCE_1, EXT_SOURCE_2, EXT_SOURCE_3) using a ridge regression model.
(3) Use binary features (only 0 and 1 variables) in the main table and fit the label using a linear regression model.

4. One-to-Many Subtable Features

In this competition, due to the presence of multiple one-to-many subtables, they cannot be directly merged into the main table. To handle these subtable data, various methods are used to extract features from these subtables and merge the extracted features into the main table. Among them, aggregation statistics, subtable meta-features, and subtable meta-features based on time series models are used. These methods help better utilize subtable data to improve the model's prediction performance.[5]

(1) Aggregation Statistics

For these auxiliary tables, aggregation grouping is performed on the SK_ID_CURR column (representing the unique identifier column for samples) to calculate statistical information for other numerical columns. The statistical information includes mean, sum, maximum, minimum, median, variance, and weighted mean (assigning greater weight to more recent data). In addition, subsets of the auxiliary tables are selected by setting certain conditions, and aggregate statistical features are calculated for the subset data. The subset selection conditions include setting a time range (e.g., data from the last three years) and conditional filtering (e.g., historically approved/rejected application samples).

---

[5] The reference material can be found at https://www.kaggle.com/competitions/home-credit-def ault-risk/discussion/64503.

(2) Auxiliary Table Meta-features

The one-to-many auxiliary table meta-feature involves passing the main table labels to the auxiliary table, training a model to generate predictions, and then aggregating and returning the predictions to the main table. For the construction logic, refer to the content of Sect. 2.3.6 "Constructing Auxiliary Table Meta-features." In this case, such features were constructed for the previous_application table (user historical loan application record table) and the bureau table (user loan application record table with other financial institutions). Additionally, the POS_CASH_balance table (user loan monthly balance snapshot table) is aggregated on the SK_ID_CURR column (representing the unique identifier column for samples) and months to calculate aggregate statistical information for other columns, thereby constructing a new one-to-many auxiliary table, and then generating auxiliary table meta-features for this table.

(3) Time Series Model-Based Auxiliary Table Meta-features

One-to-many auxiliary table time series meta-features refer to constructing auxiliary table meta-features using time series models when there is time series information in the auxiliary table data, which can consider the chronological order of samples. For the construction logic, refer to the content of Sect. 2.3.6 "Construct Auxiliary Table Meta-features Using Time Series Models." In this data, most of the one-to-many auxiliary tables contain time information of samples (e.g., the installment repayment data table contains the repayment time for each sample). In this case, the GRU model was used to construct time series meta-features for the following four auxiliary tables: bureau table (user loan application record table with other financial institutions), credit_card_balance table (credit card monthly balance snapshot table), installments_payments table (installment repayment record table), and POS_CASH_balance table (user loan monthly balance snapshot table).

3.3.2 Model

1. Downsampling Strategy

Due to the imbalance of the dataset in the competition, with positive samples (label 1) being less than one-tenth of the training set, training a model with such imbalanced data is not ideal. Therefore, the plan involved downsampling the negative samples, using one-third of the negative samples combined with the positive samples for training each time, repeating this process three times, and taking the average of the three results as the final result. The core code logic is as follows.

```
# Obtain the values corresponding to the minority and majority
classes
minority = y.value_counts().sort_values().index.values[0]
```

```
majority = y.value_counts().sort_values().index.values[1]

# Separate the X and y of the minority and majority classes
X_min = X.loc[y == minority]
y_min = y.loc[y == minority]
X_maj = X.loc[y == majority]
y_maj = y.loc[y == majority]

kf = KFold(3, shuffle=True, random_state=42)
for rest, this in kf.split(y_maj):
    X_maj_sub = X_maj.iloc[this]
    y_maj_sub = y_maj.iloc[this]

    # Merge the full minority class with one-third of the majority
class each time
    X_sub = pd.concat([X_min, X_maj_sub])
    y_sub = pd.concat([y_min, y_maj_sub])
```

2. LightGBM

The model section used LightGBM, which is high in performance and efficiency. Training was conducted using fivefold cross-validation, and the average of the five prediction results was taken as the final result. Each fold used the downsampling strategy and was executed with three different sets of model parameters.

```
y = data['TARGET']
# Perform 5-fold cross-validation
folds = StratifiedKFold(n_splits=5, shuffle=True, random_
state=90210)
oof_preds = np.zeros(data.shape[0])
sub_preds = np.zeros(test.shape[0])
feature_importance_df = pd.DataFrame()

scores = []  # Store the score of each fold
for n_fold, (trn_idx, val_idx) in enumerate(folds.split(data,
data['TARGET'])):
    trn, val = data.iloc[trn_idx], data.iloc[val_idx]

    model = LGBMClassifier(
        n_estimators=5000,
        learning_rate=0.03,
        num_leaves=26,
        metric='auc',
        colsample_bytree=0.28,
        subsample=0.95,
        max_depth=4,
        reg_alpha=4.8299,
        reg_lambda=3.6335,
        min_split_gain=0.005,
        min_child_weight=40,
```

```
            silent=True,
            verbose=-1,
            n_jobs=16,
            random_state=n_fold * 6666,
            class_weight={0: 1, 1: 1}
    )

    clf = bagging_classifier(model, 3)  # Perform downsampling

    clf.fit(trn_x, trn_y,
            eval_set=[(val_x, val_y)],
            eval_metric='auc',
            verbose=200,
            early_stopping_rounds=100,
            categorical_feature=cat_feats
            )

    oof_preds[val_idx] = clf.predict_proba(val_x)[:, 1]
    sub_preds += clf.predict_proba(test_x)[:, 1] / folds.n_
splits

    fold_score = roc_auc_score(val_y, oof_preds[val_idx])
    scores.append(fold_score)
    print('Fold %2d AUC : %.6f' % (n_fold + 1, fold_score))

    fold_importance_df = pd.DataFrame()
    fold_importance_df["feature"] = features
        fold_importance_df["importance_gain"] = clf.feature_
importances_gain_
        fold_importance_df["importance_split"] = clf.feature_
importances_split_
    fold_importance_df["fold"] = n_fold + 1
    feature_importance_df = pd.concat([feature_importance_df,
fold_importance_df], axis=0)

print('Full AUC score %.6f +- %0.4f' % (roc_auc_score(y, oof_
preds), np.std(scores)))
```

3. Feature Importance

Using the LightGBM model trained in "2. LightGBM," the feature importance of the model can be obtained, as shown in Fig. 3.8.

It can be seen that the top-ranked feature AMT_PAY_YEAR is a constructed business feature, constructed by dividing the loan amount by the loan annuity, indicating how many years are needed for repayment. Following closely are EXT_SOURCE_ 1, EXT_SOURCE_2, EXT_SOURCE_3, which are three credit scores that show strong correlation in correlation analysis. Ranked fifth is the original feature DAYS_ BIRTH, representing the customer's age in days.

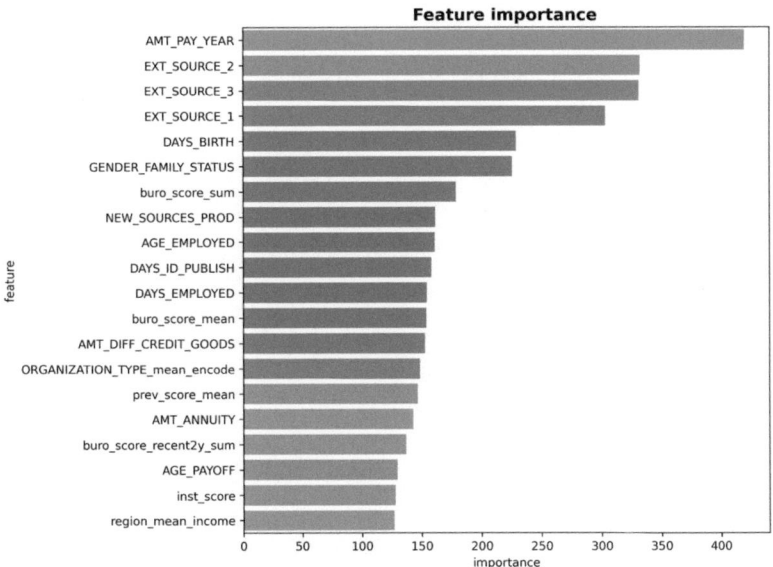

Fig. 3.8 Feature importance obtained from the LightGBM model

3.3.3   *Ensemble Learning*

The ensemble learning module adopts Bayesian weight search, with the specific approach as follows.

In the single model stage, due to the use of fivefold cross-validation, out-of-fold predictions of the training set were obtained. Different weights are assigned to each single model prediction result, and the AUC on the training set after different weight fusion is calculated. The goal is to maximize the AUC, and the search process uses the optuna package for Bayesian search, with the code as follows.

```
# Out-of-fold prediction results of the training set
all_preds = train[['sub1', 'sub2', 'sub3']].values

# Assign weights to single models with the goal of maximizing
the AUC
def max_auc(params):
    preds = None
    for index, val in enumerate(params.keys()):
        if index == 0:
            preds = params[val] * all_preds[:, 0]
        else:
            preds += params[val] * all_preds[:, index]
    param_sum = 0
    for key, val in params.items():
```

Table 3.1 Results of single models and ensemble learning

Model	Private score	Public score
lgb_1	0.80026	0.80491
lgb_2	0.80021	0.80492
lgb_3	0.80011	0.80476
ensemble	0.80028	0.80494

```
        param_sum += val
    preds = preds / param_sum
    score = roc_auc_score(train['target'], preds)
    return score

# Generate weights
def objective(trial):
    params = {}
    for i in range(3):
        params[f"w{i + 1}"] = trial.suggest_float(f'w{i + 1}',
0, 1)
    score = max_auc(params)
    return score

# Conduct the experiment
study = optuna.create_study(direction='maximize')
study.optimize(objective, n_trials=500)

# Obtain the best weights from the experiment
weights = list(study.best_params.values())
weights = [w / sum(weights) for w in weights]

# Apply the best weights to the prediction results of the test
set
final_pred = None
for i, model in enumerate(['sub1', 'sub2', 'sub3']):
    if i == 0:
        final_pred = test[model] * weights[i]
    else:
        final_pred += test[model] * weights[i]
```

The scores of each single model and ensemble learning are shown in Table 3.1. The final ensemble result scored 0.80028 on the private leaderboard (ranked 36th out of 7176).

Chapter 4
Natural Language Processing: Theoretical Part

The main purpose of natural language processing (NLP) is to study various theories and methods that enable effective communication between humans and computers using natural language. With the help of NLP technology, computers can understand and generate human language, allowing people to use computers more conveniently without mastering complex and variable programming languages. A partial framework diagram of NLP technology is shown in Fig. 4.1.

Natural language processing technology is widely applied, not only in the currently popular field of text generation represented by ChatGPT but also in traditional areas such as text classification, intelligent customer service, and key entity extraction. Since many tasks are open-ended and evaluation standards are difficult to define accurately, such as story continuation and text summarization, most current NLP competitions still focus on classic tasks like classification and regression. For example, on Kaggle, from 2018 to 2022, there were 16 competitions primarily focused on NLP tasks in the industrial competitions (featured competition), as shown in Table 4.1. Among the top ten solutions for each competition, 11 out of 16 were mainly focused on text classification and regression, accounting for a large proportion.

To quickly get started with NLP competitions, in addition to understanding the domain background, it is also necessary to understand the current mainstream technologies. NLP-related problems can be solved through conventional feature engineering + machine learning methods, or through current deep learning methods based on pre-trained language models and fine-tuning. Generally, the former has higher processing efficiency, but the effect is not as good as the latter.

Compared to processing efficiency, most NLP competitions, mainly on Kaggle, place more emphasis on model accuracy. Therefore, modeling with pre-trained language models as the core is the mainstream solution for NLP tasks at present. Based on the input–output forms of pre-trained language models, mainly BERT, the current mainstream natural language processing competitions are divided into three

© Tsinghua University Press 2026
K. Xu, *Data Mining Competition Practices*,
https://doi.org/10.1007/978-981-95-3446-3_4

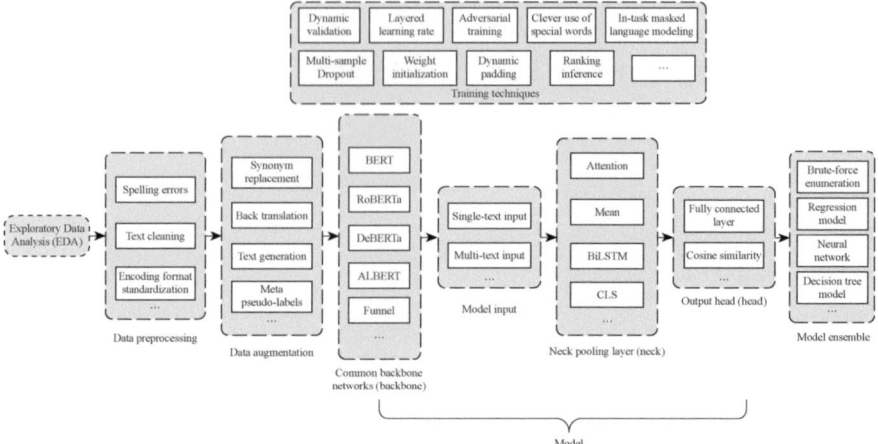

Fig. 4.1 Partial framework diagram of NLP technology

categories: text classification, text regression, and question answering based on the original text.

Among them, different categories can be further subdivided according to scenarios (input–output forms), such as text classification tasks with token-level output, which are similar to named entity recognition, and text regression tasks with two text inputs and a single regression output, which are similar to text similarity calculation.

No matter how complex and variable the task itself is, essentially, from the perspective of the modeling process, all tasks are the same. The factors that lead to differences in model accuracy are mainly concentrated on the selection of the most appropriate input–output definition method, training techniques, and various detailed modeling methods according to the characteristics of the task and dataset.

Therefore, taking Fig. 4.1 as an example, the following will gradually introduce the different stages of modeling from the perspective of a general process and then analyze the specific competition task scenarios suitable for various detailed modeling methods in different stages.

4.1   Exploratory Data Analysis

Compared to structured data, exploratory data analysis in NLP revolves around text, focusing on key information such as word count and high-frequency words. The operations involved are very simple and can be achieved by calling the APIs encapsulated in libraries like Matplotlib, pandas, WordCloud, and Seaborn with just a few lines of code.

Table 4.1 Summary of NLP-related industrial competitions hosted on the Kaggle platform (2018–2022)

Time	Competition	Main approach
2017.12.19–2018.3.20	Toxic Comment Classification Challenge	Text classification
2018.11.7–2019.2.14	Quora Insincere Questions Classification	Text classification
2019.3.30–2019.7.19	Jigsaw Unintended Bias in Toxicity Classification	Text regression
2019.10.29–2020.1.23	TensorFlow 2.0 Question Answering	Question answering based on original text
2019.11.23—2020.2.10	Google QUEST Q&A Labeling	Classification
2020.3.24—2020.6.22	Jigsaw Multilingual Toxic Comment Classification	Classification
2020.3.24—2020.6.16	Tweet Sentiment Extraction	Classification
2021.3.24—2021.6.22	Coleridge Initiative-Show US the Data	Named entity recognition
2021.5.4—2021.8.2	CommonLit Readability Prize	Regression
2021.11.9—2022.2.7	Jigsaw Rate Severity of Toxic Comments	Regression
2021.12.15—2022.4.16	Feedback Prize-Evaluating Student Writing	Named entity recognition
2022.2.2—2022.5.4	NBME-Score Clinical Patient Notes	Question answering based on original text
2022.3.22—2022.6.21	U.S. Patent Phrase to Phrase Matching	Text matching/regression
2022.5.12—2022.8.5	Google AI4Code –Understand Code in Python Notebooks	Others
2022.5.25—2022.8.24	Feedback Prize-Predicting Effective Arguments	Categorization
2022.8.31—2022.11.29	Feedback Prize-English Language Learning	Regression

4.1.1 Text Word Count Statistics

Text word count refers to the number of words in each piece of text in the dataset after tokenization. Text word count is an important parameter in NLP tasks; overly long texts may cause model encoding anomalies or insufficient running resources, while overly short texts may affect the training effect of the model. Therefore, it is necessary to perform text truncation or choose an appropriate model for processing based on specific situations (detailed analysis on how to choose based on actual situations will be provided later). Below is a code example for text word count statistics.

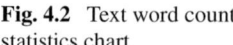

Fig. 4.2 Text word count statistics chart

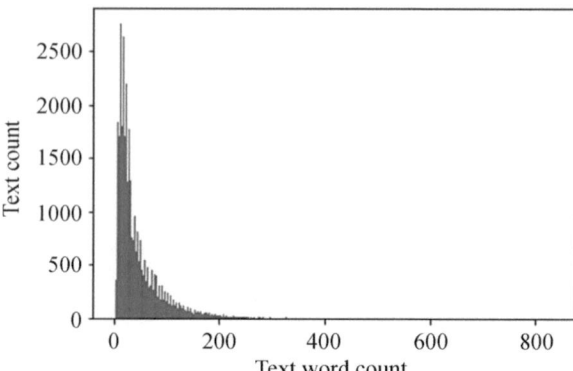

```
import pandas as pd
import matplotlib.pyplot as plt
import seaborn as sns
train_df       =        pd.read_csv('../input/feedback-prize-
effectiveness/train.csv')
train_df['word_count']   =   train_df.text.apply(lambda   x:
len(x.split()))
# Tokenize each text using spaces and count the number of words.
For Chinese, jieba can be used for tokenization
sns.histplot(data=train_df, x="word_count")
plt.show()
```

The visualization result of the text word count statistics is shown in Fig. 4.2.

4.1.2 High-Frequency Word Statistics

High-frequency word statistics refer to counting the words that appear frequently in a corpus. This is usually done to understand the characteristics of the corpus or to prepare for subsequent information extraction and specific processing tasks. The following code displays high-frequency words in the text in the form of a word cloud, where the more frequently a word appears, the larger its the font in the word cloud effect.

```
import pandas as pd
import matplotlib.pyplot as plt
from wordcloud import WordCloud
train_df       =        pd.read_csv('../input/feedback-prize-
effectiveness/train.csv')
texts_list = train_df['text'].to_list()
```

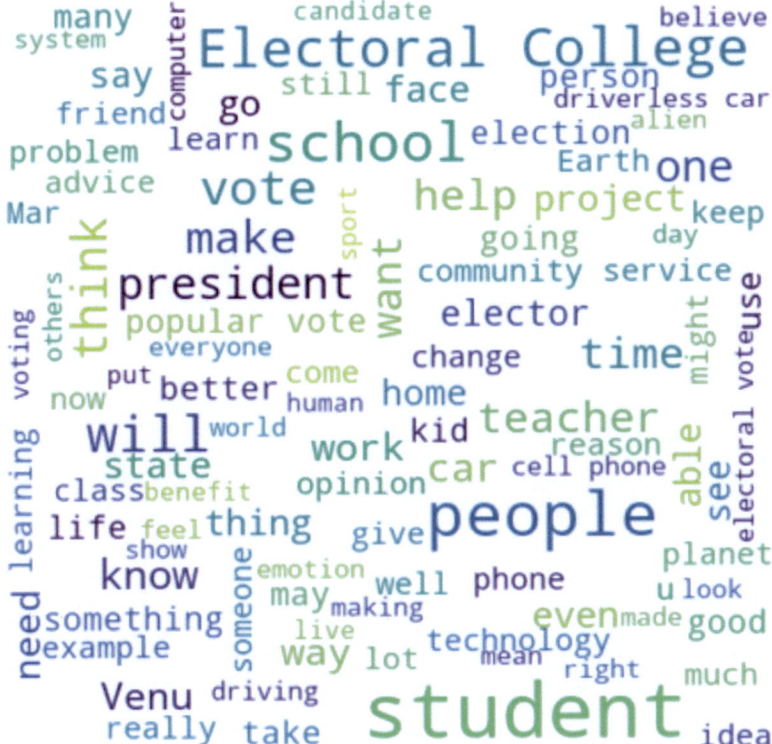

Fig. 4.3 High-frequency word statistics via word cloud

```
texts = ' '.join(texts_list)
wordcloud   =   WordCloud(max_font_size=50,   max_words=100,
width=500,
height=500, background_color="white").generate(texts)
plt.imshow(wordcloud, interpolation="bilinear")
plt.axis("off")
```

The visualization result is shown in Fig. 4.3.

4.2 Data Preprocessing

Text data preprocessing is primarily aimed at eliminating minor errors in the text or standardizing the text format to make it more "readable," thereby facilitating model understanding and improving the model's performance.

1. Spelling correction

Spelling correction refers to correcting input errors or common word usage errors in
the text. Since there is often a certain proportion of spelling errors in the text, which
can mislead the model, ignoring these errors may affect the text processing results.
For these subtle yet common errors, open-source tools like Neuspell can be used for
batch correction. An example of usage is as follows.

```
import neuspell
from neuspell import BertChecker
import pandas as pd
df_train = pd.read_csv('train.csv')
checker = BertChecker()
checker.from_pretrained() # Download the model weight files
from Google Cloud and load them
df_train['corrected_text'] = checker.correct_strings(df_
train.text.values) # Perform error correction
```

2. Text cleaning

Text cleaning usually refers to using regular expressions and BeautifulSoup (a Python
library that can extract data from HTML or XML files) to remove special symbols,
emoji, URLs, HTML tags, and other text that is unhelpful or even negatively impacts
the task. Text cleaning can reduce the interference of abnormal text on modeling
results. However, in some tasks, special symbols are key to improving results, such as
newline characters for text segmentation tasks and exclamation marks for sentiment
classification tasks, so it is necessary to decide whether to clean a certain type of text
based on the specific situation.
    Below is a code example of a general cleaning method.

```
import re
from bs4 import BeautifulSoup
def text_cleaning(text):
    '''
    Perform the following operations in order:
    (1) Remove embedded URLs.
    (2) Remove HTML tags.
    (3) Remove emojis.
    (4) Remove special symbols such as &, #, etc.
    (5) Remove extra spaces.
    '''
    template = re.compile(r'https?://\S+|www\.\S+') # Remove
embedded URLs
    text = template.sub(r"", text)
    soup = BeautifulSoup(text, 'lxml')  # Remove HTML tags
```

```
        only_text = soup.get_text()
        text = only_text
        emoji_pattern = re.compile("["
                                        u"\U0001F600-\U0001F64F"
                                        u"\U0001F300-\U0001F5FF"
                                        u"\U0001F680-\U0001F6FF"
                                        u"\U0001F1E0-\U0001F1FF"
                                        u"\U00002702-\U000027B0"
                                        u"\U000024C2-\U0001F251"
                                        "]+", flags=re.UNICODE)
        text = emoji_pattern.sub(r", text)  # Remove emojis
        text = re.sub(r"[^a-zA-Z\d]", " ", text)  # Remove special
symbols
        text = re.sub(' +', ' ', text)                  # Remove extra
spaces
        text = text.strip()  # Remove leading and trailing spaces
        return text
```

3. Encoding format unification

Encoding format unification refers to handling garbled and abnormal characters. It can improve the quality of the text, thereby enhancing the model's prediction performance.

Below is a code example of a general encoding format unification method.

```
import codecs
from typing import Optional, Tuple
from text_unidecode import unidecode
import pandas as pd
def resolve_encodings_and_normalize(text: str) -> str:
    """Resolve encoding issues and uniformly handle abnormal
characters."""
    text = (
        text.encode("raw_unicode_escape")
            .decode("utf-8", errors="replace_decoding_with_
cp1252")
        .encode("cp1252", errors="replace_encoding_with_utf8")
            .decode("utf-8", errors="replace_decoding_with_
cp1252")
    )
    text = unidecode(text)
    return text
def  replace_encoding_with_utf8(error:  UnicodeError)  ->
Tuple[bytes, int]:
    return error.object[error.start: error.end].encode("utf-
8"), error.end
def  replace_decoding_with_cp1252(error:  UnicodeError)  ->
Tuple[str, int]:
```

```
                          return      error.object[error.start:
error.end].decode("cp1252"), error.end
# Register encoding and decoding error handlers for 'utf-8' and
'cp1252'
codecs.register_error("replace_encoding_with_utf8",
replace_encoding_with_utf8)
codecs.register_error("replace_decoding_with_cp1252",
replace_decoding_with_cp1252)
df_train = pd.read_csv('train.csv')
train_df['text'] = [resolve_encodings_and_normalize(i) for i
in train['text']]
```

4.3   Data Augmentation

Data augmentation refers to expanding the data using external knowledge or models based on labeled data to increase the data volume. Obtaining sufficient labeled data is always a challenge in NLP competitions, where it is common to encounter situations with only a few thousand labeled data points, which is far from enough for training models like BERT. Therefore, this section introduces several commonly used text data augmentation methods and implements them using the text augmentation tool nlpaug.

4.3.1   Synonym Replacement

As the name suggests, synonym replacement involves replacing some words in the text with other words that have the same meaning. This method can introduce some additional knowledge from the perspective of words, but not all additional knowledge is helpful for specific tasks, so it can only improve the training effect of the model under certain situations.

The usage example provided by nlpaug is as follows, where the replacement model used in the example is WordNet, and other models such as PPDB can also be selected.

```
import nlpaug.augmenter.word as naw
text = 'The quick brown fox jumped over the lazy dog'
aug = naw.SynonymAug(aug_src='wordnet')
augmented_text = aug.augment(text)
```

4.3.2 Back Translation

Back translation refers to translating a text from language A to language B and then back to language A. Assuming the original text is in English, first specify a translation language different from the original, such as German, translate the original text into German, and then translate the German text back into English.

This method can leverage the model knowledge from excellent open-source translation platforms or the translation model knowledge from the open-source community, using an intermediate language as a bridge to expand the entire text. Back translation can introduce differences between languages, but these differences are not always helpful for the task, and the translation model itself also has accuracy issues, so while introducing additional knowledge, errors are also introduced. Therefore, the choice of translation language and translation model is not fixed and needs to be adjusted based on actual results.

The usage example provided by nlpaug is as follows, where the text data language is English, the intermediate language is German, and the open-source translation model from the Hugging Face community "facebook/wmt19-en-de" is used.

```
import nlpaug.augmenter.word as naw
text = 'The quick brown fox jumped over the lazy dog'
back_translation_aug = naw.BackTranslationAug(
    from_model_name='facebook/wmt19-en-de',
    to_model_name='facebook/wmt19-de-en'
)
back_translation_aug.augment(text)
```

4.3.3 Text Generation

Text generation in data augmentation usually refers to using generation models or summarization models to continue or summarize the original text to obtain new text. It is equivalent to introducing knowledge by introducing differences between different task forms. Similar to other text augmentation methods, it improves the effect by introducing additional knowledge. Therefore, it is also necessary to choose whether to use generation or summarization and the corresponding task model based on actual results.

1. Generation

The usage example provided by nlpaug is as follows, where "xlnet-base-cased" is a model provided by the Hugging Face open-source community that can be used for text generation, and n=3 represents the number of new texts generated from one text is 3.

```
import nlpaug.augmenter.sentence as nas
aug        =        nas.ContextualWordEmbsForSentenceAug(model_
path='xlnet-base-cased')
augmented_texts = aug.augment(text, n=3) # n represents the
number of new texts generated
```

2. Summarization

The usage example provided by nlpaug is as follows, where "t5-base" is a model
provided by the Hugging Face open-source community that can be used for text
summarization tasks.

```
import nlpaug.augmenter.sentence as nas
article = """
The history of natural language processing (NLP) generally
started in the 1950s, although work can be
found from earlier periods.
......
Little further research in machine translation was conducted
until the late 1980s when the first statistical machine trans-
lation systems were developed.
"""
aug = nas.AbstSummAug(model_path='t5-base')
augmented_text = aug.augment(article)
```

4.3.4   Meta Pseudo Labels

Meta Pseudo Labels is a method proposed in the paper Meta Pseudo Labels [1],
which involves training a teacher model on a labeled dataset. The teacher model
generates pseudo-labeled data on an unlabeled dataset for the student model to learn
from. This method can increase the data volume while ensuring that the knowledge
in the augmented data is more aligned with the task. It is generally more stable and
effective than previous augmentation methods and is suitable for situations where
both labeled and unlabeled data are available. In simple terms, Meta Pseudo Labels
uses labeled data as a benchmark to continuously optimize the model on unlabeled
data. A schematic diagram of Meta Pseudo Labels is shown in Fig. 4.4.
    The steps for Meta Pseudo Labels are as follows.

1. First, train the "student" on labeled data and select the best-performing "student"
    as the "teacher."
2. Use the "teacher" to predict and assign pseudo labels on unlabeled data.
3. Train a new "student" with the pseudo-labeled data.

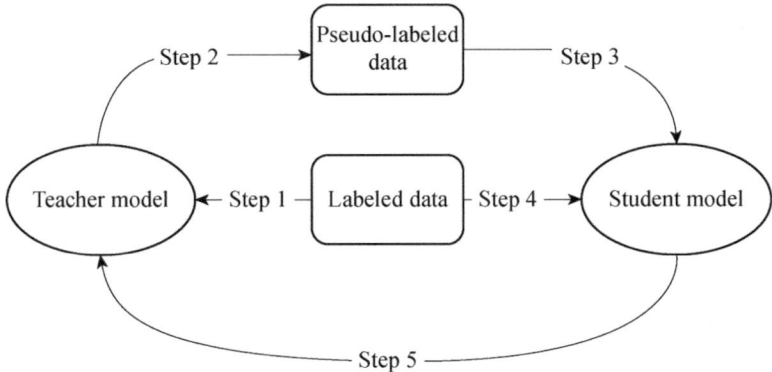

Fig. 4.4 Schematic diagram of Meta Pseudo Labels

4. Validate with labeled data.
5. Select the "student" with the best validation performance as the new "teacher."
6. Repeat steps (2)–(5) iteratively until the optimization limit is reached.

Additionally, to prevent the "student" from prematurely knowing the validation data, the concept of out-of-fold (oof) is used, which involves dividing the labeled data into n equal parts, using $n - 1$ parts as the data the student needs to learn from, and the remaining 1 part to validate the student's learning effect. The $n - 1$ parts correspond to the labeled data in step (1), and the remaining 1 part corresponds to the labeled data used for validation in step (3).

4.4 Model

In the era where attention mechanisms are shining, as one of the representative works of pre-trained models, the evergreen tree of NLP competitions—the BERT series models—is an unavoidable topic for every contestant. Due to its well-supported open-source community, most tasks can use the API provided by open-source libraries to quickly call BERT series models for training and prediction. Therefore, this section focuses on the common usage ideas of BERT series models in most competitions, combined with the well-known open-source NLP library—transformers provided by Hugging Face, and some DIY tips for specific tasks.

4.4.1 NLP Competition All-Purpose—BERT

An introduction to the BERT model is shown in Fig. 4.5.

Below is a brief introduction to the characteristics, components, and corresponding functions of the BERT model.

What is BERT? Simply put, BERT is a deep neural network that can understand text. To elaborate, it is a deep neural network that can understand the context of text and has a certain level of language understanding ability after being pre-trained on large-scale corpora.

So how does BERT transform text into a mathematical expression that a computer can understand? This involves BERT's tokenizer. The tokenizer first splits the sentence into subwords that exist in the vocabulary. Typically, it adds a CLS token as a start symbol and a SEP token as a separator at the beginning and end of the sentence. Then, it replaces the split subwords with their unique IDs in the vocabulary, resulting in a representation similar to a one-dimensional array, known as input ids, completing the transformation from text to mathematical expression. Additionally, different models have outputs like attention masks, token type IDs, and offset

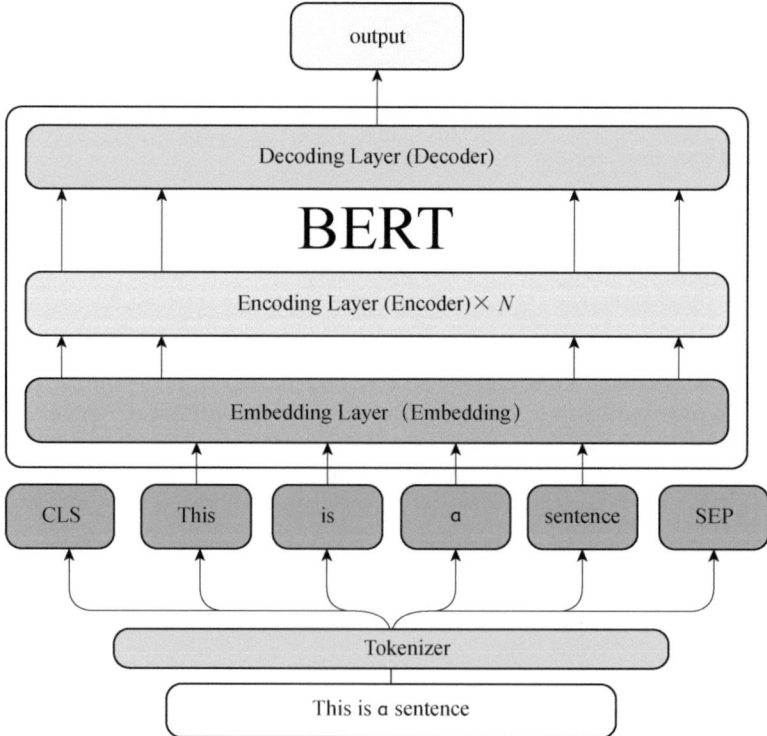

Fig. 4.5 Introduction to the BERT model

mappings during the tokenization phase, and some normalization operations, such as converting to lowercase and removing special symbols.

After obtaining the mathematical expression, what operations does the BERT model perform internally? It can be simply summarized as embedding layer feature expression, encoding layer feature extraction (Transformer Encoder), and output layer output acquisition (Transformer Decoder). The embedding layer can be likened to a feature dictionary that stores the feature vectors of each subword, responsible for mapping the one-dimensional tensor (converted from the array) obtained by the model to two dimensions, obtaining a richer feature expression space, and then handing it over to the encoding layer for deep feature extraction. The output layer, responsible for converting the encoding layer results into task outputs, needs to be designed according to specific tasks, so usually, we only use the pre-trained weights of the embedding and encoding layers of the pre-trained model.

The above is a brief introduction to the BERT series of models. Next, we will discuss the design ideas for the input/output parts in the context of specific tasks using the BERT model, as well as how to select an appropriate pre-trained model.

4.4.2 Common Model Backbones and Their Characteristics

Although many excellent derivative models have been proposed after BERT, each model has its advantages, disadvantages, and specific applications. Some models may not perform well on the General Language Understanding Evaluation, a multi-task natural language understanding benchmark and analysis platform created by institutions such as New York University and the University of Washington (GLUE) leaderboard. However, because the pre-training corpus happens to be quite similar to the text in the task scenario, they can achieve better convergence results. Therefore, a brief introduction to several commonly used models is provided below, along with some tips for choosing pre-trained models in different scenarios.

1. RoBERTa

The characteristics of this model are as follows:

(1) Better performance and robustness than BERT: RoBERTa is an enhanced version of BERT, with major improvements focused on training optimization (computing power, data volume, batch size), as well as some training method improvements (removal of the "next sentence prediction" pre-training task, dynamic masking, byte-level text encoding).
(2) More comprehensive support from the open-source community: Due to RoBERTa's good robustness and its impressive performance on many tasks, it has become quite popular after its introduction. Therefore, most tasks can be applied using the open-source transformer library.
(3) More derivative open-source models: The Hugging Face model library offers up to 6230 RoBERTa models pre-trained for different tasks and languages.

(4) Absolute position encoding: Does not support processing single inputs exceeding a specified length.

2. DeBERTa

The characteristics of this model are as follows:

(1) Currently at the top of the GLUE leaderboard: DeBERTa is currently the state of the art (SOTA) for most natural language understanding (NLU) tasks, with many optimal solutions using DeBERTa and its variants.
(2) A considerable number of derivative open-source models: The open-source community currently offers nearly 600 open-source DeBERTa pre-trained models covering multiple languages and tasks.
(3) Relative position encoding: Supports processing single inputs exceeding a specified length.

3. ALBERT

The characteristics of this model are as follows:

(1) Lightweight: ALBERT is a lightweight improved version of BERT, reducing the number of parameters while ensuring performance.
(2) The open-source community offers a considerable number of derivative open-source models.
(3) ALBERT uses absolute position encoding, which supports processing single inputs up to a specified length.

4. Funnel-Transformer

The characteristics of this model are as follows:

(1) Higher processing efficiency: Adopts a funnel-shaped structure, compressing along the "sequence length" dimension as the network deepens, and finally using a decoder for upsampling to ensure consistent dimensions, overall reducing the number of parameters and improving processing efficiency.
(2) Funnel-Transformer uses relative position encoding to handle sequences more effectively.

5. Model selection tips

In addition to the models mentioned above, there are many open-source models not covered here, each with its own characteristics. So how should one choose in a competition? Here are some main ideas:

(1) Maximum text word count: As mentioned in Sect. 2.1, if most of the text word counts in a task dataset exceed 514, which is the limit for most absolute position encoding models, it is preferable to consider models with relative position encoding. If you want to use models with absolute position encoding, you can refer to the solution provided in Sect. 6.11—LSG.

(2) Competition evaluation criteria (time or performance): If the competition imposes a short online inference time limit, such as within an hour, it is necessary to choose the model with the best single-model performance, such as the DeBERTa series. If the competition focuses more on overall performance and time is ample, different types of pre-trained models can be integrated.

(3) Text data language: Different competitions may involve text data in different languages. You can choose the corresponding pre-trained model based on the specific language. If multiple languages are involved, you can choose a multilingual pre-trained model.

(4) Domain-specific tasks: In some competitions, the text data may be strongly related to a specific domain, such as patent similarity matching, which includes many specialized terms and abbreviations. Sometimes, the model may not "recognize" frequently appearing specialized terms, leading to poor performance. Therefore, choosing a pre-trained language model related to the domain can sometimes result in single-model performance that surpasses the integrated model performance.

6. Basic invocation methods and common APIs

After introducing the model characteristics, we briefly introduce the method of introducing pre-trained models with code, as well as the APIs for saving and loading model weights.

(1) Method of introducing pre-trained models. Thanks to the powerful open-source library transformers, after finding the required model name on the Hugging Face community, you can import it with just one line of code.

```
from transformers import AutoModel
model    =    AutoModel.from_pretrained('funnel-transformer/
large')
```

(2) API for saving model weights.

```
from transformers import AutoModel
import torch
model    =    AutoModel.from_pretrained('funnel-transformer/
large') # Load model structure
torch.save(model.state_dict(), "saved_model.pth")
```

(3) API for loading model weights.

```
from transformers import AutoModel
import torch
model    =    AutoModel.from_pretrained('funnel-transformer/
large') # Load model structure
state       =        torch.load('saved_model.pth',        map_
location=torch.device('cuda'))
model.load_state_dict(state)
```

4.4.3 Designing Inputs for BERT-like Models

First, clarify a question: What mainly constitutes the input of BERT-like models? As introduced in Sect. 4.1, BERT-like models use tokenizers to convert text into mathematical expressions that the model can understand, i.e., subwords and special tokens after text segmentation, and then obtain the corresponding IDs.

"Flowing text, ironclad special tokens" is a description of the characteristics of BERT-like model inputs. Therefore, designing inputs for different task scenarios can be likened to designing embedding templates for special tokens for different tasks. Next, we will introduce common special token templates for different tasks using the Hugging Face transformers library.

1. Single-text input tasks (text classification, named entity recognition, text scoring, etc.)

As the name suggests, single-text input tasks refer to tasks where each input is a complete text, as shown in the examples below.

(1) Malicious comment detection (text classification), predicting each comment text to output a malicious level, such as severe, moderate, mild, or non-malicious.
(2) Text auto segmentation (named entity recognition), segmenting an argumentative article to output the segment category of each subword, such as introduction, body, conclusion, etc.
(3) Readability scoring (text scoring), scoring the summary of an article to output the readability score of the article.

These tasks, apart from text input, do not require additional information to be introduced and usually adopt the default embedding template, i.e., "[CLS] + tokenized subwords + [SEP]" input method, as shown in the code below.

```
import pandas as pd
from transformers import AutoTokenizer
test_df = pd.read_csv("test.csv")
text = test_df.text.values[0]# "This is a sentence"
tokenizer   =   AutoTokenizer.from_pretrained('bert-base-
uncased')
encoded_text_ids = tokenizer.encode(text)['input_ids'] #
Directly get the ID form of the tokenized input[101, 2023,
2003, 1037, 6251, 102]
tokenized_text = tokenizer.convert_ids_to_tokens(encoded_
text_ids)
['[CLS]', 'this', 'is', 'a', 'sentence', '[SEP]']
```

APIs like tokenizer.__call__, tokenizer.encode, tokenizer.batch_encode_plus can directly convert inputs into ID form. If you need to view the intermediate state of subwords + special tokens, you can use tokenizer.convert_ids_to_tokens to convert the ID form result.

2. Multiple text input tasks (original text Q&A, simple text matching, etc.)

Unlike single-text input tasks, multiple text input tasks usually require processing more than one text simultaneously, as shown in the examples below.

(1) Patient case description keyword extraction (original text Q&A), given a case text (content) and a disease description, output one or more symptom words (sentences) related to the disease in the original case text.
(2) Patent similarity prediction (simple text matching), given a classification keyword of a specific field patent and key phrases of two patents, output the similarity of the two patents in that field.

The commonality of inputs for these tasks is that there are multiple different text inputs simultaneously, usually adopting the "[CLS] + text1 + [SEP] + text2 + [SEP] + textn + [SEP]" embedding template. For cases with two inputs simultaneously, you can directly call the Tokenizer's built-in encoding API, as most model tokenizers have implementations for embedding two text inputs, as shown in the code below.

```
import pandas as pd
from transformers import AutoTokenizer
test_df = pd.read_csv("test.csv")
text1 = test_df.text.values[0] # "This is sentence1"
text2 = test_df.text.values[1] # "This is sentence2"
tokenizer   =   AutoTokenizer.from_pretrained('bert-base-
uncased')
encoded_text = tokenizer(text1, text2,
                    add_special_tokens=True) # {'input_ids':
[101, 2023, 2003, 6251, 2487, 102, 2023, 2003, 6251, 2475, 102],
'token_type_ids': [0, 0, 0, 0, 0, 0, 1, 1, 1, 1, 1], 'attention_
mask': [1]}
```

```
tokenized_text = tokenizer.convert_ids_to_tokens( encoded_
text ['input_ids']) # ['[CLS]', 'this', 'is', 'sentence',
'##1', '[SEP]',' this', 'is', 'sentence',
'##2','[SEP]']
```

In the case of having more than two text inputs simultaneously, you need to first concatenate multiple texts with a separator and then call the encoding API of the Tokenizer for processing. The code is as follows.

```
import pandas as pd
from transformers import AutoTokenizer
test_df = pd.read_csv("test.csv")
text1 = test_df.text.values[0] # "This is sentence1"
text2 = test_df.text.values[1] # "This is sentence2"
textn = test_df.text.values[n] # "This is sentenceN"
tokenizer    =    AutoTokenizer.from_pretrained('bert-base-
uncased')
combined_text = text1 + tokenizer.sep_token + text2 +
tokenizer.sep_token + textn # "This is sentence1[SEP]This
is sentence2[SEP]This is sentenceN"
encoded_text = tokenizer(combined_text) # {'input_ids':
[101, 2023, 2003, 6251, 2487, 102, 2023, 2003, 6251, 2475, 102,
2023, 2003, 6251, 2078, 102], 'token_type_ids': [0, 0, 0, 0,
0, 0, 0, 0, 0, 0, 0, 0, 0, 0, 0, 0], 'attention_mask': [1]}
tokenized_text = tokenizer.convert_ids_to_tokens( encoded_
text['input_ids']) #['[CLS]', 'this', 'is', 'sentence',
'##1', '[SEP]', 'this', 'is', 'sentence', '##2', '[SEP]',
'this', 'is', 'sentence', '##n', '[SEP]']
```

3. Other task forms (structured text classification, etc.)

In addition to designing model inputs for the above two types of tasks, it is also necessary to consider the input information corresponding to many other tasks. Some input information not only contains text but also includes additional elements, such as argumentative essay paragraph review (structured text classification), where given an article, the start and end character position indices of different paragraphs, and the paragraph name of each paragraph (introduction, body, conclusion, etc.), the writing situation of each paragraph is output (excellent, average, poor). It is necessary to consider not only the differences in content between different paragraphs but also the differences in the positions of different paragraphs. This more complex task does not have a universal input paradigm, but essentially still uses special tokens for processing, which will be further introduce in Sect. 6.4.

4.4.4 *Designing the Neck of BERT-like Models*

As mentioned in the introduction to the BERT model, under normal circumstances, we only use the embedding layer and encoding layer pre-trained weights of the pre-trained model, while the decoding layer of the model needs to be designed for specific tasks. The decoding layer includes the neck and the head, corresponding to the feature extraction method and forward propagation method in the custom model structure, respectively.

Generally, the neck only performs secondary feature processing on the output obtained from the model's encoding layer, while the head is responsible for converting the output into the format required by the task, further processing the neck's output to obtain the final output.

As a transitional layer, the model neck needs to handle the feature output of the encoding layer on the one hand and cooperate with the model head to obtain the final task output on the other. Therefore, the following two points generally need to be considered when designing.

(1) Transition of output form: According to the task form, convert the form of the feature output from the encoding layer to cooperate with the model head to obtain the output form that best fits the task requirements. For example, for a question-answering task based on the original text, it is necessary to obtain the position of the answer in the original text from the text dimension, so it is necessary to retain the sequence length(seq_len) dimension in the feature output of the encoding layer.

(2) Impact on effect: Different necks will also have an impact on the actual effect, but the specific effect is closely related to the pre-trained model itself, the task scenario, and the data. If the best effect is pursued, it is necessary to try various neck designs as much as possible while ensuring point(1).

The sample code for implementing a custom model class is as follows.

```python
class Model(nn.Module):
        def __init__(self, cfg, config_path=None,
pretrained=False):
        super().__init__()
        self.cfg = cfg
        self.config = AutoConfig.from_pretrained(cfg.model,
output_hidden_ states=True)
        self.model = AutoModel.from_pretrained(cfg.model,
config=self. config)
        self.neck = AttentionPool(self.config.hidden_size)
        self.fc = nn.Linear(self.config.hidden_size, 6)
    def feature(self, inputs):
        outputs = self.model(**inputs)
        feature = self.neck(outputs.last_hidden_state,
inputs['attention_ mask'])
        return feature
```

```
def forward(self, inputs):
    feature = self.feature(inputs)
    output = self.fc(feature)
    return output
```

Below, we mainly introduce four types of necks: attention, average pooling, BiLSTM, and CLS pooling.

1. Attention

This usually involves calculating the global feature weights at dim = 1 (seq_len) based on the hidden_dim dimension and weighting them to obtain a two-dimensional tensor output of the form batch_size × hidden_dim. An example implementation class is as follows.

```
class AttentionPool(nn.Module):
    def __init__(self, in_dim):
        super().__init__()
        self.attention = nn.Sequential(
        nn.Linear(in_dim, in_dim),
        nn.LayerNorm(in_dim),
        nn.GELU(),
        nn.Linear(in_dim, 1),
        )
    def forward(self, x, mask):
        w = self.attention(x).float()
        w[mask==0]=float('-inf')
        w = torch.softmax(w,1)
        x = torch.sum(w * x, dim=1)
        return x
```

2   Average pooling

This usually refers to obtaining the output of the last encoding layer and directly averaging over the sequence length(seq_len) dimension to get a two-dimensional tensor of the form batch_size × hidden_dim. An example implementation class is as follows.

```
class MeanPooling(nn.Module):
    def __init__(self):
        super(MeanPooling, self).__init__()
    def forward(self, last_hidden_state, attention_mask):
            input_mask_expanded = attention_mask.unsqueeze(-
1).expand
(last_hidden_state.size()).float()
```

```
        sum_embeddings=torch.sum(last_hidden_state * input_
mask_expanded,1)
            sum_mask = input_mask_expanded.sum(1)
            sum_mask = torch.clamp(sum_mask, min=1e-9)
            mean_embeddings = sum_embeddings / sum_mask
            return mean_embeddings
```

3. BiLSTM

```
class BiLSTMPool(nn.Module):
    def __init__(self, hidden_size, bilstm_hidden_size):
        super().__init__()
      self.bilstm = nn.LSTM(hidden_size, bilstm_hidden_size,
bidirectional= True, batch_first=True)
    def forward(self, last_hidden_state):
        neck_output, _ = self.bilstm(last_hidden_state)
        return neck_output
```

4. CLS pooling

This usually refers to obtaining the output of the last encoding layer and taking the first feature vector of the sequence length(seq_len) dimension to get an output of the form batch_size × hidden_dim. An example implementation class is as follows.

```
class CLSPooling(nn.Module):
    def __init__(self):
        super(CLSPooling, self).__init__()
    def forward(self, last_hidden_state):
        neck_output = last_hidden_state[:,0,:]
        return neck_output
```

4.4.5 Designing the Output of BERT-like Models

After obtaining the output from the neck, for different task output forms, we briefly introduce several output (head) design methods and their applicable tasks.

1. Single fully connected layer

Suitable for most tasks. It is usually used to directly reduce the hidden dimension(hidden_dim) to the target dimension target_size. If the neck output is batch_ size × seq_len × hidden_size, the output is of the form batch_size × seq_len × target_size; if the neck output is batch_size × hidden_size, the output is of the form batch_size × target_size. The former is suitable for tasks like original text question answering that need to retain the sequence length(seq_len) dimension to obtain the corresponding position words and sentences, while the latter is suitable for text classification/regression tasks.

An example model class implementation for the original text question-answering task is as follows.

```python
class Model(nn.Module):
    def __init__(self, cfg):
        super().__init__()
        self.cfg = cfg
        self.config = AutoConfig.from_pretrained(cfg.model_p,
output_hidden_states=True)
        self.model = AutoModel.from_pretrained(cfg.model_p,
        self.bilstm_hidden_size = 768
        self.neck = BiLSTMPool(self.config.hidden_size,
self.bilstm_hidden_size)
        self.head = nn.Linear(self.bilstm_hidden_size*2, 1)
    def feature(self, inputs):
        outputs = self.model(**inputs)
        last_hidden_states = outputs[0]
        return self.neck(last_hidden_states)
    def forward(self, inputs):
        neck_feature = self.feature(inputs)
        output = self.head(neck_feature)
        return output
```

The example implementation of the model class for text classification and regression tasks is as follows.

```python
class Model(nn.Module):
    def __init__(self, cfg):
        super().__init__()
        self.cfg = cfg
        self.config = AutoConfig.from_pretrained(cfg.model_
p, output_
hidden_states=True)
        self.model = AutoModel.from_pretrained(cfg.model_p,
        self.neck = CLSPooling()
        self.head = nn.Linear(self.config.hidden_size, 1)
    def feature(self, inputs):
        outputs = self.model(**inputs)
        last_hidden_states = outputs[0]
        return self.neck(last_hidden_states)
    def forward(self, inputs):
```

```
            neck_feature = self.feature(inputs)
            output = self.head(neck_feature)
            return output
```

2. Cosine similarity

This usually refers to calculating the similarity between two outputs processed by the neck from the model's two encoding layers, resulting in a batch_size × score output. Cosine similarity is suitable for text matching tasks using a Siamese network or dual-tower structure.

The implementation code for the example model class is as follows.

```
class Model(nn.Module):
    def __init__(self, cfg):
        super().__init__()
        self.cfg = cfg
        #anchor
        self.anchor_config = AutoConfig.from_
pretrained(cfg.anchor_model, output_hidden_states=True)
        self.anchor_model = AutoModel.from_
pretrained(cfg.anchor_model, config=self.anchor_config)
        self.anchor_neck = AttentionPool(self.anchor_
config.hidden_size)
        #target
        self.target_config = AutoConfig.from_
pretrained(cfg.target_model, output_hidden_states=True)
        self.target_model = AutoModel.from_
pretrained(cfg.target_model, config=self.target_config)
        self.target_neck = AttentionPool(self.target_
config.hidden_size)
    def anchor_feature(self, inputs):
        last_hidden_state = self.anchor_model(**inputs)[0]
        neck_output = self.anchor_neck(last_hidden_state)
        return output
    def target_feature(self, inputs):
        last_hidden_state = self.target_model(**inputs)[0]
        neck_output = self.target_neck(last_hidden_state)
        return eck_output
    def forward(self, anchor_inputs,target_inputs):
        anchor_feature = self.fc_dropout(self.anchor_
feature(anchor_
inputs))
        target_feature = self.fc_dropout(self.target_
feature(target_
inputs))
        # Use torch's cosine similarity function to calculate
the
similarity between two encoded tensors
```

```
                        output  =  torch.cosine_similarity(anchor_
    feature,target_feature,
    dim=1)
            return output
```

3. Others

The model head essentially serves the task output, and besides the two relatively simple and common methods listed above, there are many other design approaches for single-task scenarios. For example, in Sect. 6.4, we will combine the use of special tokens with head design to output structured text classification results containing multiple segments to be predicted.

4.5 Model Ensemble

As the saying goes, "Three cobblers with their wits combined equal Zhuge Liang, " and model ensemble is just like that. It combines multiple pre-trained trained models together, using specific methods to achieve a comprehensive application of these models on test data. The purpose of this approach is to combine the strengths of each model to achieve "synergistic effects" and improve the final result.

For text classification and text regression problems, you can use direct weighted averaging of the results, determine different models' linear weights based on out-of-fold (OOF), and construct stacking through multi-layer models.

For question answering based on the original text, the output is usually directly predicted using a "pointer" method to obtain the start and end positions of the answer. For the fusion of pointer outputs, it often refers to the idea of object detection tasks, such as weighted boxes fusion (WBF). However, note that different models may have different tokenizers, meaning the same word in the same text may be located at different indices by different models. Therefore, the index representing the same word in different models' outputs may be different, so you can locate the word-level pointer to the most basic char-level before performing the fusion operation.

For named entity recognition, the output is usually obtained by classifying each word into its respective category, where consecutive words of the same category form the same entity. For the fusion of word classification outputs, the category probabilities of all words in the entire text can be uniformly viewed as a probability matrix with the shape (seq_len, target_num). Then, weighted fusion and other operations can be performed on the probability matrix. However, the fusion of word classification outputs also faces an index alignment issue similar to pointer output fusion, so the probability matrix needs to be aligned before performing fusion operations.

4.6   Training Techniques

This section will introduce common training techniques used during competitions and also supplement some details from previous content.

4.6.1   Dynamic Validation

Dynamic validation usually refers to using a backward parameter update as a trigger for a validation step during model training. After the number of steps reaches a certain level, validation data is used to obtain the current model's validation set accuracy. If it is the current global optimum, the model weights are saved, and the closer the model accuracy is to the expected value, the fewer the interval steps between two validations.

This method essentially refines the original fixed validation granularity of one epoch validation (the process of validating all samples in the training dataset) by sacrificing training efficiency to improve model performance. It is generally used when model training parameters are determined and extreme performance is pursued. It is not recommended to use it in the early stages of a competition.[1]

1. Predefine validation intervals for different accuracies

The code is as follows.

```
schedule=[(float('inf'), 400),(0.450, 10),(0.445, 2), (0, 0)]
```

The code includes four intervals, representing:

(1) When the validation loss is less than infinity (i.e., before the loss is calculated), the validation interval is 400 steps;
(2) When the validation loss is less than 0.45, the validation interval is 10 steps;
(3) When the validation loss is less than 0.445, the validation interval is 2 steps;
(4) When the validation loss is less than 0, an exception is thrown.

2. Define the dynamic validation implementation class

The code is as follows.

```
class EvaluationScheduler:
        def __init__(self, evaluation_schedule, penalize_
factor=1,
max_penalty=8):
        self.evaluation_schedule = evaluation_schedule
```

---

[1] The link is https://www.kaggle.com/code/chamecall/clrp-finetune-roberta-large.

```python
        self.evaluation_interval = self.evaluation_
schedule[0][1]
        self.last_evaluation_step = 0
        self.prev_loss = float('inf')
        self.penalize_factor = penalize_factor
        self.penalty = 0
        self.prev_interval = -1
        self.max_penalty = max_penalty
    def step(self, step):
        # If the current training step is greater than or equal to
the sum of the last evaluation step and the current evaluation
interval, perform an evaluation and set the last evaluation
step to the current step.
        if step >= self.last_evaluation_step + self.evaluation_
interval:
            self.last_evaluation_step = step
            return True
        else:
            return False
    def update_evaluation_interval(self, last_loss):
        # Use the input validation loss as the previous valida-
tion loss. Compare the validation losses in the dynamic evalu-
ation preset intervals from left to right, and find the preset
interval that is closest to but greater than the validation loss
as the current evaluation interval.
        cur_interval = -1
        for i, (loss, interval) in enumerate(self.evaluation_
schedule[:-1]):
            if self.evaluation_schedule[i + 1][0] < last_loss <
loss:
                self.evaluation_interval = interval
                cur_interval = i
                break
        if last_loss > self.prev_loss and self.prev_interval
== cur_interval:
                self.penalty += self.penalize_factor
                self.penalty = min(self.penalty, self.max_
penalty)
                self.evaluation_interval += self.penalty
        else:
                self.penalty = 0
        self.prev_loss = last_loss
        self.prev_interval = cur_interval
```

3. Call during the training phase

The code is as follows.

```
def train_loop(folds, fold):
        train_folds = folds[folds['fold'] != fold].reset_
index(drop=True)
        valid_folds = folds[folds['fold'] == fold].reset_
index(drop=True)
    train_dataset = TrainDataset(CFG, train_folds)
    valid_dataset = ValidDataset(CFG, valid_folds)
    train_loader = DataLoader(train_dataset,
                    batch_size=CFG.batch_size,
                    shuffle=True,
                    num_workers=CFG.num_workers,
pin_memory=True, drop_last=True)
  valid_loader = DataLoader(valid_dataset,
                    batch_size=CFG.batch_size,
                    shuffle=True,
                    num_workers=CFG.num_workers,
pin_memory=True, drop_last=True)
        model    =    CustomModel(CFG,    config_path=None,
pretrained=True)
  optimizer = AdamW(optimizer_parameters, lr=CFG.encoder_lr,
eps=CFG. eps, betas=CFG.betas)
    criterion = nn.SmoothL1Loss(reduction='mean')
    evaluation_scheduler = EvaluationScheduler(CFG.schedule)
    best_score=np.inf
    for epoch in range(CFG.epochs):
      model.train()
      scaler = torch.cuda.amp.GradScaler(enabled=CFG.apex)
      for step, (inputs, labels) in enumerate(train_loader):
        for k, v in inputs.items():
            inputs[k] = v.to(device)
        labels = labels.to(device)
        y_preds = model(inputs)
        loss = criterion(y_preds, labels)
        scaler.scale(loss).backward()
        scaler.step(optimizer)
        scaler.update()
        optimizer.zero_grad()
# The following is the key code
```

```
# After each step is completed, it will determine whether the
step interval for validation is met. If it is met, a validation
will be carried out, and the validation loss or score will be
obtained.
        if evaluation_scheduler.step(steps):
            score, predictions = valid_fn(valid_loader, model,
criterion, device)
            if best_score > score:
                best_score = score
                torch.save(model.state_dict()," model.pth")
    # Update the evaluation interval according to the validation
    loss or score
    evaluation_scheduler.update_evaluation_interval(score)
    return True
```

4.6.2   Hierarchical Learning Rate

The hierarchical learning rate refers to setting different learning rates for different layers of the model when initializing the optimizer. In pre-trained language models, the knowledge in the layers closer to the input (shallow layers) has stronger generality, and the parameters change less during the training process. On the other hand, the knowledge in the layers closer to the output (deep layers) is more closely related to the task, and the parameters change more significantly during the training. Therefore, we can set progressive learning rates according to the layers of the model from shallow to deep, which helps the model converge better.

This method is generally used when the model has a large number of encoding layers (24 layers or more). Usually, large models are more sensitive to the learning rate, and setting progressive learning rates can improve the convergence speed and performance of large models. However, the specific hierarchical progressive strategy also needs to be continuously explored. Generally, a linear increasing method is adopted for the encoding layers, and for the neck and head parts, the learning rate of the last encoding layer is directly used.

The following is a code example.

```
def  get_optimizer_params(model,  encoder_lr,  decoder_lr,
weight_decay=0.0):
    # Get the names and parameters of each layer in the model
    named_parameters = list(model.named_parameters())
    parameters = []
    # Change the hierarchical learning rate every k layers
    increase_lr_every_k_layer = 1
```

```
    # Store different learning rates in a list. Here, 24 is the
number of encoding layers, 1 is the initial learning rate multi-
plier, and 5 is the final learning rate multiplier. The total
number of layers // k layers for change = number of changes =
number of different learning rates
    lrs = np.linspace(1, 5, 24 // increase_lr_every_k_layer)
    # A list of keywords included in the model layers without
weight decay
    no_decay = ["bias", "LayerNorm.bias", "LayerNorm.weight"]
    # Iterate over the names and parameters of each layer in the
model
        for layer_num, (name, params) in enumerate(named_
parameters):
        # If the name contains keywords of layers without weight
decay, set the weight decay to 0
        weight_decay = 0.0 if any(nd in name for nd in no_decay)
else 0.01
            # Usually, the embedding layer and sampling layer do
not need to set hierarchical learning rates. More often, hier-
archical learning rates are set for encoding layers. A model
contains multiple encoding layers, and the names of each subdi-
vided layer in the encoding layer are composed of the layer type
and the layer number of the encoding layer, separated by a "."
        splitted_name = name.split('.')
        # encoder_lr is the default learning rate and also the
initial learning rate
        lr = encoder_lr
            # Determine whether the currently iterated layer
belongs to the encoding layer and judge the current learning
rate according to the layer number of the encoding layer
        if str.isdigit(splitted_name[3]):
            layer_num = int(splitted_name[3])
            lr = lrs[layer_num // increase_lr_every_k_layer] *
encoder_lr
            print(name, lr)
        # In the model naming rules of Hugging Face, the names of
each layer of the pre - trained model generally contain "model".
To distinguish custom layers, names with "model" are usually
not used. So, to set learning rates for custom layers, a reason-
able judgment method needs to be set and distinguished from the
layers of the pre - trained model
        if 'model' not in splitted_name:
            lr = decoder_lr
            print(name, lr)
        parameters.append({"params": params,
                           "weight_decay": weight_decay,
                           "lr": lr})
        return parameters
```

4.6.3   Adversarial Training

Adversarial training is a method that introduces a certain degree of disturbance or noise during the training process. A suitable adversarial training strategy helps enhance the model's stability and its robustness when facing unknown data. Common methods include but are not limited to FGM, AWP, PGD, etc.

This method is suitable for most situations, but it also has the issues of longer time costs and higher hardware costs. Taking AWP as an example, the sample code includes two parts: implementation class sample code and training call sample code.

1.   Implementation class sample code[2]

```
class AWP:
    def __init__(
        self,
        model,
        optimizer,
        adv_param="weight",
        adv_lr=1e-5,
        adv_eps=0.01,
        adv_step=1,
        scaler=None
    ):
        self.model = model
        self.optimizer = optimizer
        self.adv_param = adv_param
        self.adv_lr = adv_lr
        self.adv_eps = adv_eps
        self.adv_step = adv_step
        self.backup = {}
        self.backup_eps = {}
        self.scaler = scaler
    def attack_backward(self, inputs, labels,criterion):
        self._save()
        for i in range(self.adv_step):
            self._attack_step()
            with torch.cuda.amp.autocast(enabled=CFG.apex):
                y_preds = self.model(inputs)
                    # Inside the implementation class of AWP,
an implementation also needs to be carried out according to the
actual input/output of the model, the calculation of the loss
function, and the way of backpropagation.
                adv_loss = criterion(y_preds, labels)
        self.optimizer.zero_grad()
        self.scaler.scale(adv_loss).backward()
        self._restore()
```

---

[2] Reference materials can be found at https://www.kaggle.com/code/wht1996/feedback-nn-train/ notebook.

```
    def _attack_step(self):
        e = 1e-6
        for name, param in self.model.named_parameters():
            if param.requires_grad and param.grad is not None
and self. adv_param in name:
                norm1 = torch.norm(param.grad)
            norm2 = torch.norm(param.data.detach())
            if norm1 != 0 and not torch.isnan(norm1):
                r_at = self.adv_lr * param.grad / (norm1 + e) *
(norm2 + e)
            param.data.add_(r_at)
            param.data = torch.min(
            torch.max(param.data, self.backup_eps[name][0]),
self.backup_eps[name][1]
            )
    def _save(self):
        for name, param in self.model.named_parameters():
            if param.requires_grad and param.grad is not None and
self. adv_param in name:
                if name not in self.backup:
                    self.backup[name] = param.data.clone()
                    grad_eps = self.adv_eps * param.abs().detach()
                    self.backup_eps[name] = (
                        self.backup[name] - grad_eps,
                        self.backup[name] + grad_eps,
                    )
    def _restore(self,):
        for name, param in self.model.named_parameters():
            if name in self.backup:
                param.data = self.backup[name]
        self.backup = {}
        self.backup_eps = {}
```

2. Example code for training invocation

```
def train_loop(folds, fold):
        train_folds  =  folds[folds['fold']  !=  fold].reset_
index(drop=True)
    train_dataset = TrainDataset(CFG, train_folds)
    train_loader = DataLoader(train_dataset,
                    batch_size=CFG.batch_size,
                    shuffle=True,
                num_workers=CFG.num_workers, pin_memory=True,
drop_last=True)
            model    =    CustomModel(CFG,    config_path=None,
pretrained=True)
    optimizer = AdamW(optimizer_parameters, lr=CFG.encoder_lr,
eps=CFG.eps, betas=CFG.betas)
```

```
      criterion = nn.SmoothL1Loss(reduction='mean')
      for epoch in range(CFG.epochs):
        model.train()
        scaler = torch.cuda.amp.GradScaler(enabled=CFG.apex)
        awp = AWP(
          model,
          optimizer,
          adv_lr=CFG.adv_lr,
          adv_eps=CFG.adv_eps,
          scaler=scaler
        )
        for step, (inputs, labels) in enumerate(train_loader):
          for k, v in inputs.items():
            inputs[k] = v.to(device)
          labels = labels.to(device)
          y_preds = model(inputs)
          loss = criterion(y_preds, labels)
          scaler.scale(loss).backward()
          # Critical step: Perform adversarial attack after the
normal calculation of validation loss and gradients.
awp.attack_backward(inputs, labels, criterion)
# Proceed with the remaining steps as normal.
          scaler.step(optimizer)
          scaler.update()
          optimizer.zero_grad()
        torch.save(model.state_dict(),"model.pth")
      return true
```

4.6.4 Using Special Tokens to Handle Complex Information

This usually refers to using special tokens in a tokenizer to introduce additional information into the text. In the model input section, we mentioned that besides single-text and multi-text inputs, there are task scenarios that include additional inputs such as positional information that cannot directly use the special tokens provided by the tokenizer. For such cases, custom special tokens can be introduced.

Taking the Kaggle competition Feedback Prize-Predicting Effective Arguments task as an example, this task involves structured text classification, requiring the rating of different paragraphs in an article. An example of task input/output is shown in Fig. 4.6.

As can be seen, each paragraph has a corresponding paragraph name and start-end positions in addition to the text itself. So how can these three be input into the model simultaneously? The most straightforward idea is to concatenate the text of each paragraph and the corresponding paragraph type using the [SEP] special token and input it into the model. A common input method for structured text classification tasks is shown in Fig. 4.7.

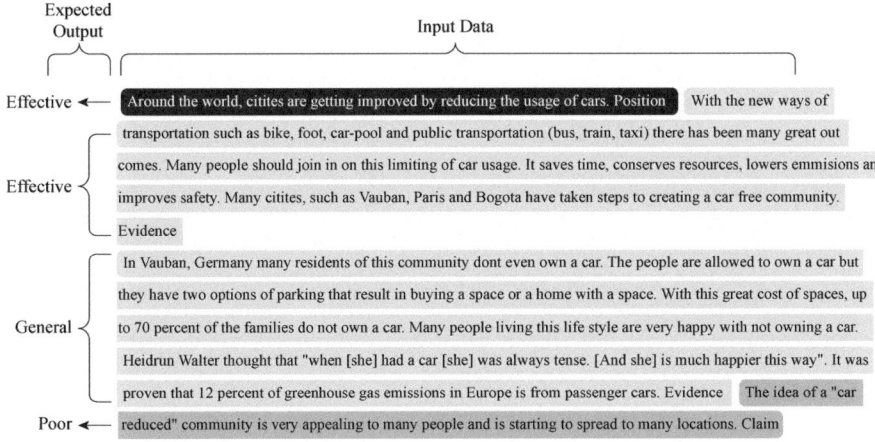

Fig. 4.6 Feedback prize-predicting effective arguments task input/output example

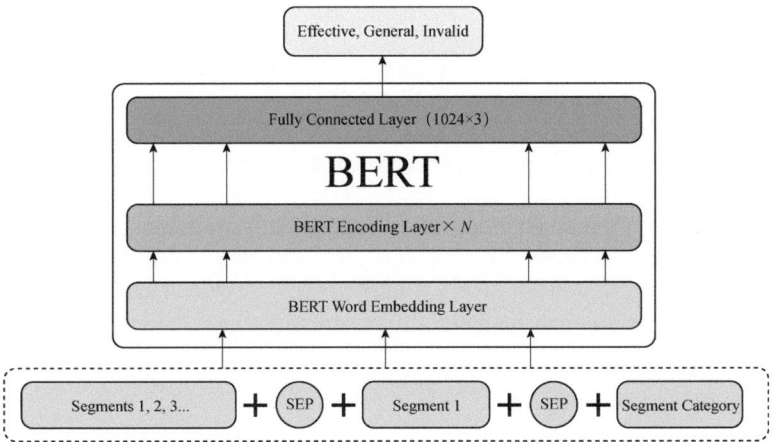

Fig. 4.7 Common input method for structured text classification tasks

This approach introduces some differences in paragraph categories and text information but ignores the relative positional differences between paragraphs, such as the fact that the opening paragraph usually comes before the closing paragraph. Moreover, this input form can only obtain the category of one paragraph at a time, while the same article has more than one paragraph that needs classification, thereby increasing time and resource costs.

So how can the positional differences between paragraphs be introduced? Two special tokens " " and " " can be defined and embedded at the start and end positions of each paragraph, respectively, while embedding the paragraph name once at each start and end position to ensure that the paragraph category difference information is

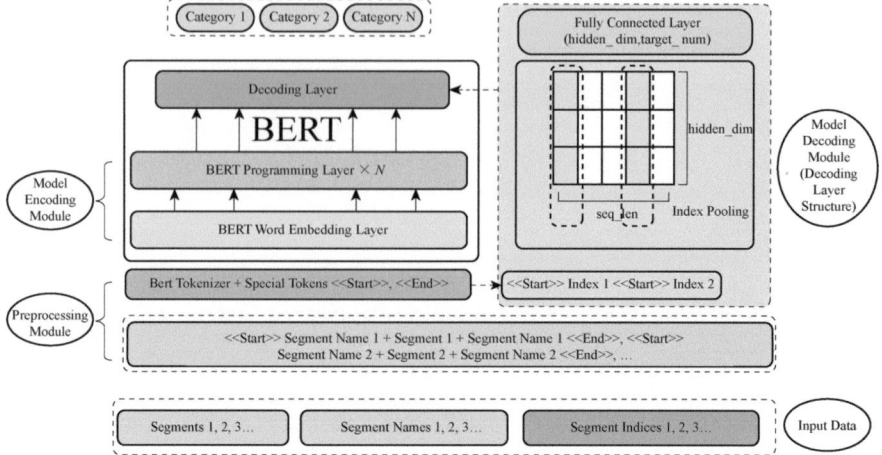

Fig. 4.8 Example of optimized input method for structured text classification tasks using special tokens

not lost. An example of the optimized input method for structured text classification tasks using special tokens is shown in Fig. 4.8.

Additionally, the position of special tokens can be utilized to design the head structure of the model, allowing for the classification results of all segments in an article to be obtained at once. Specifically, by indexing the position of the special token "<开始> " in the tokenization results, multiple feature tensors in the form of $1 \times$ hidden_dim are extracted from the seq_len dimension of the neck output. These are then reduced in dimension using a single-layer fully connected layer to obtain the category output for each segment, thus addressing the issue of time overhead.

Note: When adding special tokens, they need to be added in advance in the tokenizer. Example code is as follows.

```
from transformers import AutoTokenizer
tokenizer   =   AutoTokenizer.from_pretrained('microsoft/
deberta-v3-large')
special_tokens_dict = {'additional_special_tokens': ['<开
始>','<结束>']}
tokenizer.add_special_tokens(special_tokens_dict)
```

4.6.5 In-Task Masked Language Modeling

This usually refers to performing masked language modeling (MLM) training on a pre-trained language model using text data provided by the task scenario before fine-tuning the pre-trained model with labeled data from the task scenario. This method can improve the convergence speed and training effectiveness of the model in most cases.

This method is often used in situations where the text data volume is sufficiently large (greater than 10,000) and the length of a single text is relatively long (word count greater than 100). In tasks with small data volumes or where texts are often phrases, using MLM may not sufficiently learn inter-word relationships and other knowledge, potentially disrupting the knowledge distribution in the original pre-trained model and leading to poorer training results.

Moreover, for cases where custom special tokens are used in constructing model input/output, in-task masked language modeling can also help the model pre-learn the basic associations between custom special tokens and context, improving the effect of subsequent fine-tuning. Example code is as follows.[3]

```
import pandas as pd
import warnings
warnings.filterwarnings('ignore')
from transformers import (AutoModel, AutoModelForMaskedLM,
                AutoTokenizer, LineByLineTextDataset,
                DataCollatorForLanguageModeling,
                Trainer, TrainingArguments)
data = pd.read_csv('train.csv')
# The LineByLineTextDataset class in the transformers library
splits the whole txt text based on "\n". So, each piece of text
in the training data should be stored in a txt file with "\n" as
the separator.
text = '\n'.join(data.text.tolist())
with open('text.txt', 'w') as f:
    f.write(text)
# Set the pre-trained model for in-task masked language
modeling.
model_name = 'roberta-base'
model = AutoModelForMaskedLM.from_pretrained(model_name)
tokenizer = AutoTokenizer.from_pretrained(model_name)
tokenizer.save_pretrained('./roberta-base')
train_dataset = LineByLineTextDataset(
    tokenizer=tokenizer,
    file_path="text.txt", # Specify the text file for training
here.
    block_size=256 # block_size refers to the maximum number of
tokens in a single piece of text.
)
```

---

[3] The link is https://www.kaggle.com/code/maunish/clrp-pytorch-roberta-pretrain.

```
# mlm_probability refers to the proportion of randomly masked
tokens, usually set to 0.15.
data_collator = DataCollatorForLanguageModeling(
   tokenizer=tokenizer, mlm=True, mlm_probability=0.15
)
training_args = TrainingArguments(
   output_dir="./mlm_roberta-base",
   overwrite_output_dir=True,
   num_train_epochs=2,
   per_device_train_batch_size=16,
   evaluation_strategy='no',
   save_total_limit=2,
   load_best_model_at_end=True,
   prediction_loss_only=True,
   report_to="none"
)
trainer = Trainer(
   model=model,
   args=training_args,
   data_collator=data_collator,
   train_dataset=train_dataset
)
trainer.train()
trainer.save_model(f'./mlm-roberta-base')
```

4.6.6   Multi-sample Dropout

Multi-sample dropout [2] typically refers to applying dropout with different dropout probabilities to the neck output of the model, obtaining n outputs, and then performing n head calculations to obtain n output results, followed by averaging these results. After calculating the loss for each result, the final loss is obtained by averaging them. Simply put, it is similar to expanding a batch by n times. The comparison between original dropout and multi-sample dropout is shown in Fig. 4.9.

   This method can usually improve the convergence speed and training effectiveness of the model, but it will increase the overhead of training resources. Since dropout is generally not suitable for regression tasks, it is not recommended to use this method in regression tasks.

   The example code is as follows.

```
import torch.nn as nn
from transformers import AutoConfig, AutoModel
class Model(nn.Module):
   def __init__(self, cfg):
```

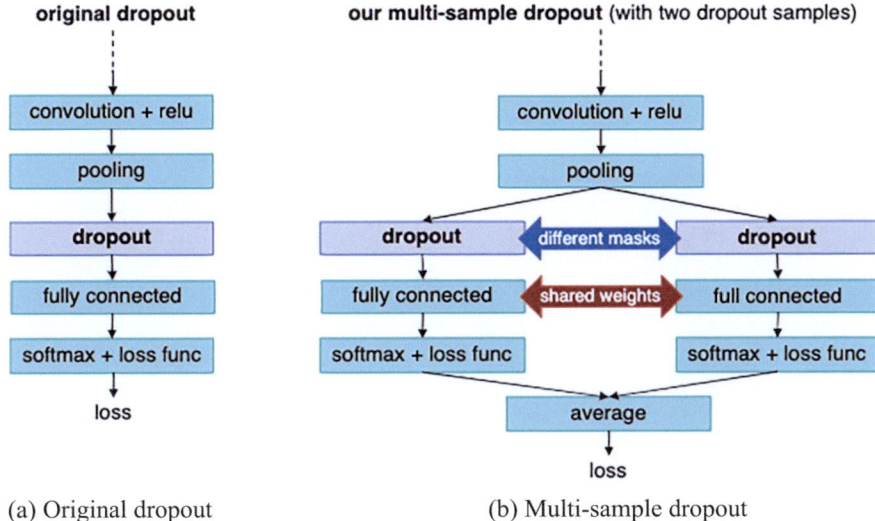

original dropout   **our multi-sample dropout** (with two dropout samples)

(a) Original dropout   (b) Multi-sample dropout

Fig. 4.9 Comparison between original dropout and multi-sample dropout

```
        super().__init__()
        self.cfg = cfg
         self.config = AutoConfig.from_pretrained(cfg.model_p,
output_hidden_states=True)
         self.model = AutoModel.from_pretrained(cfg.model_p,
config=self.config)
        self.neck = Pooling()
        self.head = nn.Linear(self.config.hidden_size, 3)
        # Set multiple different dropout rates
        self.dropout1 = nn.Dropout(0.1)
        self.dropout2 = nn.Dropout(0.2)
        self.dropout3 = nn.Dropout(0.3)
        self.dropout4 = nn.Dropout(0.4)
        self.dropout5 = nn.Dropout(0.5)
     # For better understanding, we define the loss calculation
method inside the model
    def loss(self, outputs, labels):
        loss_fct = nn.CrossEntropyLoss()
        loss = loss_fct(outputs, labels)
        return loss
    def feature(self, inputs):
        outputs = self.model(**inputs)
        last_hidden_states = outputs[0]
        return self.neck(last_hidden_states)
    def forward(self, inputs, labels):
        neck_feature = self.feature(inputs)
     # After getting the output of the neck, calculate the results
of five different dropout operations respectively
```

```
    output1 = self.head(self.dropout1(neck_feature))
    output2 = self.head(self.dropout2(neck_feature))
    output3 = self.head(self.dropout3(neck_feature))
    output4 = self.head(self.dropout4(neck_feature))
    output5 = self.head(self.dropout5(neck_feature))
    output = self.head(neck_feature)
 # Then calculate the loss for each dropout result and average
them to get the final loss
    loss1 = self.loss(output1, labels)
    loss2 = self.loss(output2, labels)
    loss3 = self.loss(output3, labels)
    loss4 = self.loss(output4, labels)
    loss5 = self.loss(output5, labels)
    loss = (loss1 + loss2 + loss3 + loss4 + loss5) / 5
    return output, loss
```

4.6.7 Model Weight Initialization

This usually refers to using the parameter initialization range preset by the pre-trained model for newly defined model structures such as neck and head, and performing parameter initialization for fully connected layers, embedding layers, and layer normalization included in the structure.

This method can usually improve the convergence speed of the model in the initial training stage and also improve the training effect. It is suitable for most situations. The only thing to note is that in the configuration files of different models, the corresponding name of the parameter initialization range may be different, generally initializer_range.

The example code is as follows.

```
class Model(nn.Module):
   def __init__(self, cfg):
      super().__init__()
      self.cfg = cfg
       self.config = AutoConfig.from_pretrained(cfg.model_p,
output_hidden_states=True)
          self.model = AutoModel.from_pretrained(cfg.model_p,
config=self.config)
      self.neck = Pooling()
      self.head = nn.Linear(self.config.hidden_size, 1)
      self._init_weights(self.neck)  # Assume the neck contains
layers
that need initialization
```

```
        self._init_weights(self.head) # In this example, the head
    is a
    fully connected layer and falls into the scope of initializa-
    tion
        # Define the method for initializing weights
        def _init_weights(self, module):
            if isinstance(module, nn.Linear):
                            module.weight.data.normal_(mean=0.0,
    std=self.config.initializer_range)
                if module.bias is not None:
                    module.bias.data.zero_()
            elif isinstance(module, nn.Embedding):
                            module.weight.data.normal_(mean=0.0,
    std=self.config.initializer_range)
                if module.padding_idx is not None:
                    module.weight.data[module.padding_idx].zero_()
            elif isinstance(module, nn.LayerNorm):
                module.bias.data.zero_()
                module.weight.data.fill_(1.0)
        def feature(self, inputs):
            outputs = self.model(**inputs)
            last_hidden_states = outputs[0]
            return self.neck(last_hidden_states)
        def forward(self, inputs):
            neck_feature = self.feature(inputs)
            output = self.head(neck_feature)
            return output
```

4.6.8 Dynamic Padding

Dynamic padding usually refers to dynamically unifying the length of text-related inputs in a batch during training, based on the maximum effective length of all text-related inputs in that batch (usually calculated using the number of 1s in the attention_mask), including input_ids, attention_mask, token_type_ids, etc.

This method can significantly save computational resources of the model and improve the speed of training and inference. Because pre-trained models need to unify the length when processing text-related inputs, the traditional method is to directly set a maximum value and pad all text-related inputs to the same length, which increases the computational load of the model. Dynamic padding, by padding according to the maximum length of each batch, can significantly alleviate this problem. Moreover, this method is suitable for most scenarios.

The sample code is as follows.

```
class DynamicPadding:
   def __init__(self, tokenizer, max_length=None):
      self.tokenizer = tokenizer
      self.max_length = max_length
   def __call__(self, inputs):
    max_length = max(sum(_["attention_mask"]) for _ in inputs)
             max_length = min(max_length, self.max_length) if
self.max_length is not None else max_length
      output = self.tokenizer.pad(encoded_inputs=inputs,
                        max_length=max_length,
                        padding=True,
                        pad_to_multiple_of=None,
                        return_tensors="pt")
      return output
```

4.6.9 Inference by Word Count Order

This usually refers to counting the number of words in each text input before processing the text into input during the inference stage, sorting all input data by word count, and then using the model for inference.

This method can save computational resource overhead during the model inference stage and improve the efficiency of model inference. However, it is important to note that after inference, the predicted results need to be re-aligned with the original order to avoid confusion.

The sample code is as follows.

```
import pandas as pd
from transformers import AutoTokenizer
test = pd.read_csv('test.csv')
submission = pd.read_csv('sample_submission.csv')
pre_tokenizer = AutoTokenizer.from_pretrained('microsoft/
deberta-v3-large')
test['tokenize_length'] = [len(pre_tokenizer(text)['input_
ids']) for text in test['full_text'].values]
test            =            test.sort_values('tokenize_length',
ascending=True).reset_index(drop=True)
# The part of model inference is omitted here. Assume that
predictions are the inference results.
test['predictions'] = predictions
# Since the data order in test is different from that in the
sample submission, it needs to be re - mapped.
submission    =    submission.merge(test[['text_id',  'predic-
tions']], on='text_id', how='left')
```

4.6.10 Gradient Checkpointing

This usually refers to using the gradient_checkpoint method provided by the transformers library during the training phase, sacrificing some (about 20%) training efficiency to reduce the memory overhead of training.

This method is not suitable for all situations and depends on whether the transformers library supports the type of pre-trained model being used.

The sample code is as follows.

```
from torch.utils.checkpoint import checkpoint
class Model(nn.Module):
  def __init__(self, cfg):
    super().__init__()
    self.cfg = cfg
     self.config = AutoConfig.from_pretrained(cfg.model_p,
output_
hidden_states=True)
      self.model = AutoModel.from_pretrained(cfg.model_p,
self.config)
        self.model.gradient_checkpointing_enable()# Enable
gradient
checkpointing
    self.neck = Pooling()
    self.head = nn.Linear(self.config.hidden_size, 1)
  def feature(self, inputs):
    outputs = self.model(**inputs)
    last_hidden_states = outputs[0]
    return self.neck(last_hidden_states)
  def forward(self, inputs):
    neck_feature = self.feature(inputs)
    output = self.head(neck_feature)
    return output
```

4.6.11 Extend Model Input Length Limit

This usually refers to converting a pre-trained language model that originally does not support long text input into a variant that supports global sparse attention (or local sparse attention, LPA) to support custom length text input.

This method can solve the problem that some absolute position encoding models do not support processing text with a word count exceeding the specified length. However, the types of pre-trained models currently supported by open-source libraries are relatively few. Taking lsg-converter as an example, the currently supported models include RoBERTa, Albert, Bart, BARThez, BERT, CamemBERT, etc.

1. Convert the pre-trained model to the LSG version

The sample code is as follows.[4]

```
!pip install lsg-converter
from lsg_converter import LSGConverter
converter = LSGConverter(max_sequence_length=4096)
# Example1
model, tokenizer = converter.convert_from_pretrained("bert-
base-uncased")
print(type(model))
#              <class           'lsg_converter.bert.modeling_lsg_
bert.LSGBertForMaskedLM'>
```

2. Use AutoModel to load the LSG version of the model

The sample code is as follows.

```
config   =   AutoConfig.from_pretrained(model_path,trust_
remote_code=True)
model    =   AutoModel.from_config(config,trust_remote_
code=True)
```

References

1. Pham, H., Dai, Z., Xie, Q., Luong, M.-T., & Le, Q. V. (2020). Meta pseudo labels [Preprint]. arXiv. https://doi.org/10.48550/arXiv.2003.10580
2. Inoue, H. (2020). Multi-sample dropout for accelerated training and better generalization [Preprint]. arXiv. https://arxiv.org/pdf/1905.09788.pdf

[4] The link is https://github.com/ccdv-ai/convert_checkpoint_to_lsg.

Chapter 5
Natural Language Processing: Practical Part

EMNLP is a prestigious international conference in the field of natural language processing, organized by the SIGDAT group of the Association for Computational Linguistics (ACL). The conference is held annually and is ranked second globally in terms of its significant influence in the field of computational linguistics.

This chapter will focus on the 2022 EMNLP Semi-supervised and Reinforcement Learning Dialogue System Challenge, co-hosted by Tsinghua University and China Mobile (see Fig. 5.1), as a case study, to explain a practical case of information extraction competition based on dialogue data.[1]

5.1 Background of the Competition

Task-oriented dialogue (TOD) systems are hindered in their large-scale effective development due to the scarcity of labeled data. Unlabeled data are often obtained in various forms, such as human-to-human dialogues, open-domain text corpora, and unstructured knowledge documents. Various semi-supervised and reinforcement methods, including pre-training, self-training, self-supervised, weakly supervised, transfer learning for zero-shot or few-shots learning, latent-variable modeling, domain adaptation, data augmentation, and reinforcement learning, all have significant application potential.

The purpose of this competition is to use various artificial intelligence technologies to construct TOD system knowledge bases and dialogue systems. This challenge focuses on semi-supervised and reinforcement dialogue systems, not only concerning the extraction of task-related knowledge—information extraction from dialogue data—but also the construction of the dialogue system itself—building task-oriented dialogue systems in customer service scenarios. Therefore, it is divided into two

---

[1] The competition address is http://seretod.org/.

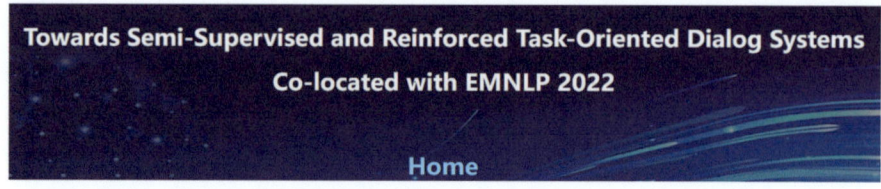

Fig. 5.1   EMNLP 2022 Semi-supervised and reinforcement dialogue system challenge

tracks: Track 1 is the information extraction task, with data sourced from customer service dialogues; Track 2 is the construction of dialogue systems (task-oriented).

This article aims to introduce Track 1 (i.e., the information extraction task based on dialogue data), which focuses on two core aspects: extracting entities and filling slot values from dialogues. In real customer service dialogue scenarios, entities may appear in various forms, and accurately identifying and extracting these entities is a crucial step in constructing a dialogue system knowledge base. Subsequently, it is necessary to further extract the slot values of entities in semantic slots by filling semantic slots.

5.2   Data Introduction

The competition provides mobile customer service dialogue (MCSD) data from China Mobile. The MCSD dataset originates from real-world dialogue scenarios, containing nearly 100,000 consultation dialogue logs between users and operator customer service after data security filtering. It is the first publicly available multi-domain task-oriented dialogue dataset of this scale, providing strong data support for research goals such as large dialogue models, colloquial human-to-human dialogue systems, and data-driven systematic dialogue analysis, and further contributing to the innovation and breakthrough of human-machine dialogue models.

The competition officially provides three parts of the dataset: 8975 training data entries, 1025 validation data entries, and 962 test data entries. The training dataset is coarsely labeled, while the validation and test datasets are finely labeled. The official ranking is based on the prediction results submitted by participants for the test set, and the test set labels are not disclosed. Each sample consists of multi-turn dialogue frames. As shown in Fig. 5.2, this is an example of a dialogue data entry in a certain round.

The labels for entity extraction are NA, business, data business, data package, package, additional package, main package, 4G package, long-distance business, and international roaming business, totaling nine categories.

The labels for slot value filling are NA, user status, business duration, data range, business cost, total data, data balance, business rules, processing channel, held package, user demand, call duration, domestic calling, account balance, domestic

```
{
  "id": "94bba9d63c097df1800482d827287e47",
  "content": [
    {
      "[SPEAKER 1]": "你好，很高兴为您服务",
      "[SPEAKER 2]": "唉你好，我想问一下，我想办那个半年_六个G那个包，我想问问那个包那个不是属于全国漫游嘛",
      "客服意图": "问候",
      "用户意图": "问候, 求助-查询（ent-1-流量范围）",
      "info": {
        "ents": [
          {
            "name": "半年_六个G那个包",
            "id": "ent-1",
            "type": "流量包",
            "pos": [
              [
                2,
                15,
                24
              ]
            ]
          }
        ],
        "triples": [
          {
            "ent-id": "ent-1",
            "ent-name": "半年_六个G那个包",
            "prop": "业务时长",
            "value": "半年",
            "pos": [
              2,
              15,
              17
            ]
          },
          {
            "ent-id": "ent-1",
            "ent-name": "半年_六个G那个包",
            "prop": "流量总量",
            "value": "_六个G",
            "pos": [
              2,
              17,
              21
            ]
          }
        ]
      }
    },
```

Fig. 5.2 Example of a dialogue data entry in the training data

receiving, out-of-package data billing, call range, account balance (arrears), affil-
iated business, out-of-package call billing, deduction date, SMS, billing method,
mutually exclusive business, data cap, totaling twenty-five categories.

5.3 Evaluation Metrics

The competition officials will evaluate the submitted models based on extraction
performance on the test set, with the evaluation metric being based on $F1$. The
evaluation score involves the following two aspects.

(1) For entity extraction, an entity is considered correctly extracted only if both the
     span annotation and entity type are correctly identified.

(2) For slot filling, correct extraction is achieved if and only if the span label and slot
     value type are correctly identified, and the slot value is correctly assigned to the
     corresponding entity. The official evaluation of slot value filling performance
     is conducted by using the Hungarian algorithm to find the best match between
     extracted entities and labels. This stage requires scoring by generating triples
     (entity, slot value, slot value position). The final score for ranking is determined
     by the average F1 score of entity extraction and slot filling.

5.4   Champion Solution

In this competition, the design concept of the champion solution is shown in Fig. 5.3.
     The overall pipeline consists of four parts: entity extraction, entity coreference
resolution, slot extraction, and entity slot alignment. First, a pre-training task is
designed and pre-trained on the training and validation datasets; then, in the model
development stage, experiments are designed to select appropriate feature extraction
networks (backbones) and decoding networks (heads), using pre-trained parameters
for initialization and fine-tuning; finally, multiple models are integrated to further
improve performance.

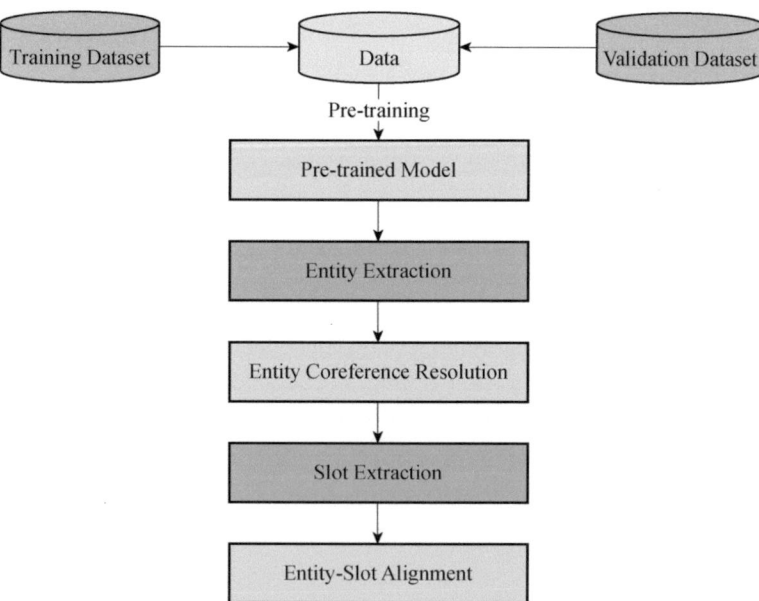

Fig. 5.3. Training process flowchart

5.4.1 Decoding Network

1. Entity extraction and slot extraction

There are many nested entities in the training data. We use the GlobalPointer network, which predicts entities in the form of start and end pointers, to indiscriminately identify both nested and non-nested entities with a "global view" [1].

For any sentence, GlobalPointer constructs an upper triangular matrix to traverse all valid spans. As shown in Fig. 5.4, each cell corresponds to an entity span. Let the length of the input sequence be n, then the number of candidate entities is $n(n + 1)/2$, and the number of actual entities is k out of $n(n + 1)/2$. Given that there are m entity types, the decoding of GlobalPointer can be simplified to a multi-label classification problem of "selecting k correct entities from $n(n + 1)/2$ candidate entities for m entity types."

The pointer design concept adopted by GlobalPointer demonstrates more reasonable characteristics compared to the traditional conditional random field (CRF) model. In practical applications, GlobalPointer does not need to recursively calculate the denominator during training as CRF does, nor does it rely on the dynamic programming algorithm during prediction. This design allows its processing to be fully parallelized, theoretically achieving a time complexity of $O(1)$ and significantly improving efficiency.

If binary classification is used in the decoding process of GlobalPointer, there will be a serious data imbalance problem. Therefore, it is converted into a multi-label classification problem with a corresponding loss function. In the entity extraction stage, there are a large number of negative samples, which can lead to relatively low model precision. To address this, we assign weight coefficients to positive and negative samples in the loss function.

```
# y_true represents the true labels, and y_pred represents the
predicted labels.
y_pred = (1 - 2 * y_true) * y_pred
```

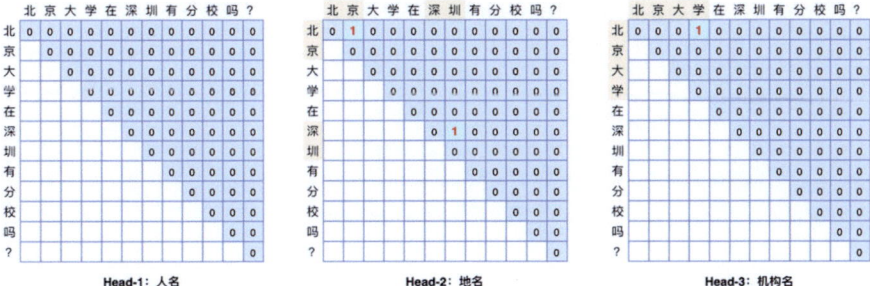

Fig. 5.4 Pointer recognition by GlobalPointer

```
y_pred_neg = y_pred - y_true * 1e12
y_pred_pos = y_pred - (1 - y_true) * 1e12
zeros = torch.zeros_like(y_pred[...,:1])
y_pred_neg = torch.cat([y_pred_neg, zeros], dim=-1)
y_pred_pos = torch.cat([y_pred_pos, zeros], dim=-1)
neg_loss = torch.logsumexp(y_pred_neg, dim=-1)
pos_loss = torch.logsumexp(y_pred_pos, dim=-1)
# The loss coefficient of negative samples is smaller than that
of positive samples.
return (0.4 * neg_loss + 0.6 * pos_loss).mean()
```

2. Entity coreference resolution

In this competition, we need to classify and aggregate the results obtained from the entity extraction in the previous stage. The process involves identifying the same entity with different expressions in the context and assigning them a unique ID. This is the process of entity coreference resolution. We adopted an end-to-end model based on deep learning, which uses word embeddings to obtain the maximum pooling and average pooling features of entities for decoding.

3. Entity slot alignment

The last module of the pipeline in this competition is the entity slot alignment task. In short, it is about assigning the extracted slot values to the corresponding entities, so as to generate the final triples (entity, slot value, slot value position) to be submitted.

First, select any two entities and slot values with the closest relative distance, and make relevant symbolic markings at the input end. Then, obtain the [CLS] feature through BERT to perform a binary classification task. The code for the symbolic marking process is as follows.

```
sorted_entities    =    sorted(entities,    key=lambda    e:
e["position"][0])
marked_text = []
curr_pos = 0
for ent in sorted_entities:
  # Slot value marking
  if ent["type"] == "triple":
    markers = ["<slot>", "</slot>"]
  # Entity marking
  elif ent["type"] == "entity":
    markers = ["<entity>", "</entity>"]
  # User attribute marking
  elif ent["type"] == "user":
    markers = ["<user>", "</user>"]
  else:
    raise ValueError()
  marked_text.extend(text[curr_pos:ent["position"][0]])
```

```
    marked_text.append(markers[0])
  marked_text.extend(text[ent["position"][0]:ent["position"]
  [1]])
    marked_text.append(markers[1])
    curr_pos = ent["position"][1]
  if text[curr_pos:]:
    marked_text.extend(text[curr_pos:])
  return "".join(marked_text)
```

During the training process, the focal loss is used to address the issue of sample imbalance. The following is the relevant code for the decoding process of training and prediction.

```
# Select the [CLS] feature of BERT
hidden_state = self.aggregation(hidden_states)
# From 768 dimensions to 2 dimensions
logits = self.cls_head(hidden_state)
# Calculate the loss
loss = None
if labels is not None:
    loss_fn = MyFocalLoss(gamma=0.5, alpha=1)
    loss = loss_fn(logits, labels)
return dict(loss=loss, logits=logits)
```

5.4.2 Feature Extraction Network

When constructing a multi-model ensemble system, selecting multiple different pre-trained models can provide richer model diversity, thereby enhancing the overall performance and robustness.

We selected the RoFormer network based on Rotary Position Embedding (RoPE).[2] Rotary Position Embedding is a design that combines with the attention mechanism to achieve "relative position encoding in the way of absolute position encoding". This model performs the best during the training process. Additionally, the following models were also selected to increase the diversity.

(1) DeBERTa[3]: It is a pre-trained model based on the self-attention mechanism. It adopts an architecture similar to BERT but introduces more improvements to the attention mechanism. It performs excellently in Chinese tasks.

---

[2] The reference link is is https://huggingface.co/junnyu/roformer_v2_chinese_char_base.

[3] The reference link is https://huggingface.co/IDEA-CCNL/Erlangshen-DeBERTa-v2-97M-Chinese.

(2) RoBERTa[4]: It is a more finely tuned version of the BERT model, trained using dynamic masking and text encoding methods.
(3) MacBERT[5]: By masking with similar words, it reduces the gap between the pre-training and fine-tuning stages, which has been proven to be effective for downstream tasks.
(4) NEZHA[6]: It improves the BERT model by using functional relative position encoding, whole-word masking, and mixed-precision training.

5.4.3 Masked Pre-training

Masked pre-training is a pre-training method for natural language processing tasks, aimed at enabling the model to understand and generate context relationships in text. Some words in the input text sequence are randomly masked or replaced with special placeholders, such as "[MASK]". The model's task is to predict the original values of these masked words. Such a prediction task forces the model to understand the semantics and grammar in the context to correctly fill in the masked words.

The advantages of the masked language model are as follows:

- Context understanding: The masked prediction task forces the model to understand and model the context relationships in the text.
- Transfer learning: The general representations obtained through pre-training can be transferred to multiple downstream tasks, eliminating the need to train from scratch.
- Performance improvement: It has achieved significant performance improvements in many natural language processing tasks, especially in semantic understanding and language generation.

We used the masked language model, an upstream task of BERT, for pre-training, with a masking rate of 0.3. The base models were selected for the pre-training process. The number of pre-training epochs was set to 5, the learning rate was set to $2e^{-5}$, the optimizer was AdamW, and the cross-entropy was used to calculate the relevant loss.

The code for the data collator that randomly masks text data is as follows:

```
# Data collator for randomly masking text data
data_collator          =          DataCollatorForLanguageMod-
eling(tokenizer=tokenizer,          mlm_probability=args.mlm_
probability)
```

[4] The reference link is https://huggingface.co/hfl/chinese -roberta-wwm-ext.
[5] The reference link is https://huggingface.co /hfl/chinese-macbert-base.
[6] The reference link is https://huggingface.co/sijunhe/nezha -base-wwm.

5.4.4 Training Techniques

1. *k*fold cross-validation

We divide the training data into four folds for cross-validation and use out-of-fold predictions to evaluate the model's generalization performance. In the resampling cross-validation process, out-of-fold predictions involve predicting the test set in each fold. This method ensures that each sample in the training dataset is predicted at least once. Specifically, the prediction results generated when each fold is used as a test set are collected to form a comprehensive list. This list accumulates all the prediction results of samples used as test sets. After training and predicting with models from all folds, the overall model accuracy can be evaluated using this summary list. The advantage of this approach is that there is enough validation data, which better highlights the model's generalization performance.

The process of kfold cross-validation is as follows.

(1) Divide the data into (approximately) equal k parts, each part is called a fold.
(2) Train a series of models, with each fold being used as a test set to evaluate accuracy, while the others are used as a training set to train the model.

2. Use of Precisely Labeled Data

We found significant class distribution differences between the training data and validation data, and the training data has poor labeling quality, while the test data and validation data have similar distributions. Based on these two reasons, it is necessary to include this batch of precisely labeled validation data in the training process to enhance the model's generalization and set data weights during training.

3. Introduction of Contextual Information

There is a lot of information scarcity in the training data, as shown in Figure 5.5, which is a training example. Only by adding contextual information can it be clarified whether the extracted entity "二十八的" represents "套餐" or "4G 套餐". Therefore, we associate this with BERT's self-attention mechanism and attempt to concatenate the entire round of dialogue content with the current dialogue information as input to enhance the interpretation of each character.

Since the current dialogue content and global context are concatenated as input, all characters are sent into BERT with an attention_mask. However, the global context information is only used to enhance content and does not participate in calculations, so we mask the context information to calculate effective loss, as shown in Fig. 5.6.

4. Decomposing Context by Characters

Since the official data converts colloquial speech information into text, there are many colloquial expressions and many pinyin expressions related to place names and special terms. BERT's Chinese tokenizer uses English rules for pinyin tokenization, resulting in many word roots, and the decomposed words lose the original meaning of the pinyin. Therefore, the context is first split into individual characters and then

```
"[SPEAKER 1]": "您好，很高兴为您服务",
"[SPEAKER 2]": "嗯喂就是，我上次叫你们给我 改的套餐是十八块钱的怎么 嗯又变成二十八的了",
"客服意图": "问候",
"用户意图": "求助-查询,提供信息",
"info": {
  {
    "name": "二十八的",
    "id": "ent-2",
    "type": "4G套餐",
    "pos": [
     [
       2,
       31,
       35|
     ]
    ]
  }
}
```

Fig. 5.5 Training data example one

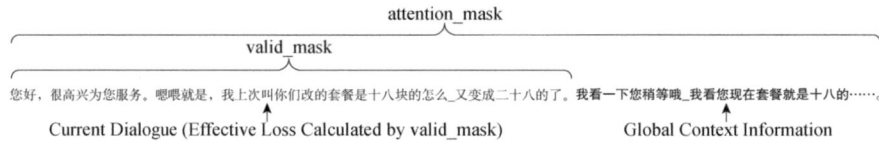

Fig. 5.6 Training data example two

sent to the BERT tokenizer. As shown in Fig. 5.7, the dialogue content contains the pinyin expression "gansu" for the place name "甘肃," which is important for the model to understand the context information. We decompose it into individual characters and send it to the BERT tokenizer.

5.   Optimizer and Learning Rate Selection

(1)   Entity extraction and slot filling

   During training, first load the decoding weights from the pre-training stage, use the Adam optimizer, select multi-label cross-entropy as the loss function, initialize the learning rate to 2e-5, and adopt a cosine annealing learning rate adjustment strategy, training for a total of five epochs.

```
"[SPEAKER 1]": "啊，好的，我明白了，就是说，嗯，给您打电话来说叫您改成那个全球通那个卡，是吧",
"[SPEAKER 2]": "我不知道，[那-gansu-local]，让我改，改着，呃 ，套餐，",

"[SPEAKER 1]":'啊',' ',' ','好','的',' ',' ','我','明','白','了',' ',' ','就','是','说',' ',' ','嗯',' ',' ','给','您','打'
' ','电','话','来','说','叫','您','改','成','那','个','全','球','通','那','个','卡',' ',' ','是','吧'

"[SPEAKER 2]":'我',' ','不',' ','知',' ','道',' ',' ','[',' ','那',' ','-',' ','g','a','n','s','u','-','l','o','c','a','l',']',' ',' ',
'让','我','改',' ',' ','改','着',' ',' ','呃',' ',' ',' ','套','餐',' ',' '
```

Fig. 5.7 Decomposing dialogue content into characters

The training code is as follows.

```
optimizer        =        torch.optim.Adam(model.parameters(),
lr=CFG.learning_rate)
# Maximum number of iterations
T_max = 500
# Minimum learning rate
min_lr = 1e-6
scheduler    =    torch.optim.lr_scheduler.CosineAnnealingLR
(optimizer, T_max=T_max, eta_min=min_lr)
```

(2) Entity coreference resolution and entity slot alignment tasks

During training, first load the decoding weights from the pre-training stage. Use the AdamW optimizer with an initial learning rate set to $4e^{-5}$. Adopt a hierarchical learning rate, which means setting different learning rates for the BERT layer and the linear layer respectively. Train for a total of ten epochs and adopt a learning rate warmup strategy. The relevant code is as follows:

```
no_decay = ["bias", "LayerNorm.bias", "LayerNorm.weight"]
optimizer_grouped_parameters = [
    {'params': [p for n, p in train_model.encoder.named_
parameters() if not any(nd in n for nd in no_decay)],
    'lr': args.learning_rate, 'weight_decay': 0.01},
    {'params': [p for n, p in train_model.encoder.named_
parameters() if any(nd in n for nd in no_decay)],
    'lr': args.learning_rate, 'weight_decay': 0.0},
   {'params': [p for n, p in train_model.named_parameters() if
"bert" not in n],
    'lr': 2e-4, 'weight_decay': 0.0}
]
# Optimizer
optimizer       =       AdamW(optimizer_grouped_parameters,
lr=args.learning_rate, eps=args.min_num)
# Linear warm-up learning
scheduler = get_linear_schedule_with_warmup(
    optimizer, num_warmup_steps=args.warmup * t_total, num_
training_steps=t_total
)
```

5.4.5   *Model Ensemble*

1. Model ensemble evaluation method

Use out-of-fold prediction to evaluate the generalization performance of the model and construct an ensemble model. Aggregate the predictions of each model into a list, which contains a summary of the reserved data used as the test set during each group's training. After all the models are trained, perform a weighted average on this list to obtain a single accuracy score.

2. GlobalPointer probability fusion

The GlobalPointer decoding stage generates a four-dimensional matrix with specific dimensions of batch $\times$ type_num $\times$ L $\times$ L, where type_num is the number of categories and L is the maximum length of characters during the training process. Set the length of L to 256 or 384. If all the predicted batches are integrated, a four-dimensional matrix that occupies a large amount of memory will be obtained, and this problem will be exacerbated during the model fusion process.

Through research, it is found that the maximum lengths of entities in the training set during the entity extraction and slot extraction stages are 20 and 50, respectively. That is, there is a large amount of redundant information in the L $\times$ L matrix. We can truncate the matrix in a staggered manner according to the maximum entity length. Then, the 256×256 matrix will be reduced to 256×20 or 256×50, and the memory usage will be significantly reduced.

The implementation process is as follows:

Initialize a Numpy matrix of sample_num $\times$ type_num $\times$ L $\times$ max_lengh_entity, where sample_num is the number of samples in the test set and max_lengh_entity is the maximum entity length. As shown in Fig. 5.8, it is the model probability fusion and decoding process during the entity extraction stage (truncate the prediction matrix in a staggered manner according to the maximum entity length of 20 to obtain the effective probability). First, initialize a zero matrix of size sample_num $\times$ type_num $\times 256 \times 20$. Then, truncate the prediction results type_num $\times 256 \times 256$ obtained by each model in a staggered manner according to the maximum length of 20 to obtain a matrix of size sample_num $\times$ type_num $\times 256 \times 20$, and fill it into the initialized zero matrix. Finally, obtain the final result by averaging the probabilities.

The relevant code for initializing the probability matrix is as follows:

```
# Initialize the probability matrix
data = json.load(open(eval_file))
sample_num = 0
for item in tqdm(data, desc="Reading" ):
    sample_num += len(item["content"])
```

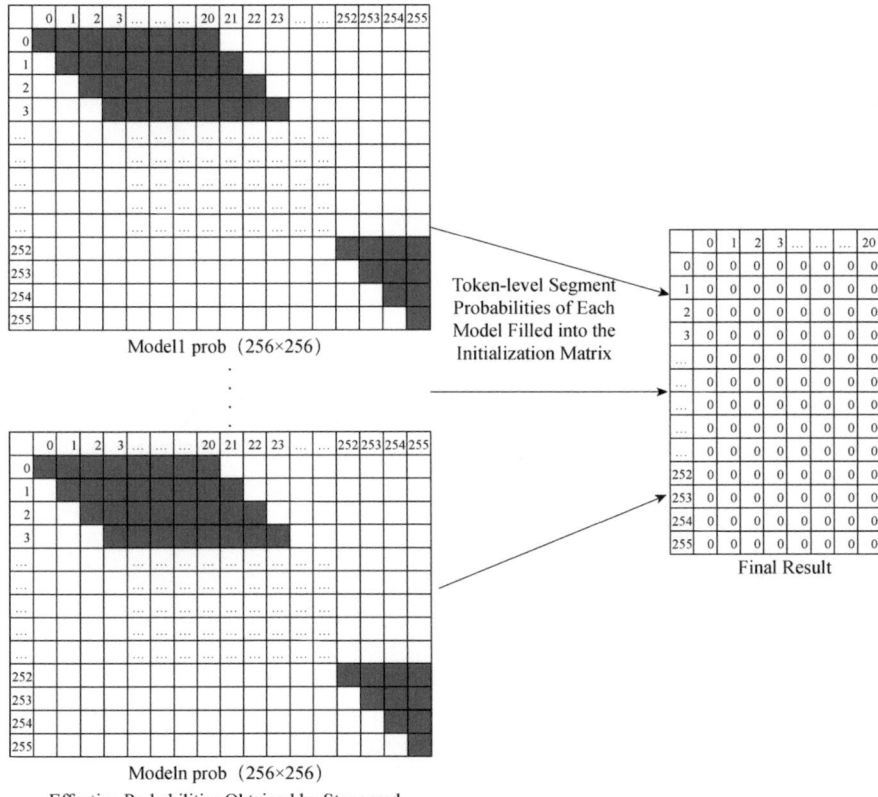

Fig. 5.8 Model probability fusion and decoding process in the entity extraction stage

```
prob_result   =   np.zeros((sample_num,ENT_CLS_NUM,256,ner_
maxlen),dtype=
np.float32)
```

The code for truncating the matrix in a staggered way and filling the initialized probability matrix is as follows.

```
# Obtain the predicted probability matrix
pred_prob = logits.sigmoid()
# Get the current batch size
sample_num = pred_prob.shape[0]
max_len = min(pred_prob.shape[2], 256)
start_index = batch_tag * batchsize
for i in range(max_len):
```

```
span = min(20, max_len - i)
# Fill the initialized probability matrix prob_result
        prob_result[start_index:start_index  +  sample_
num,   :,   i,  :span]  +=  pred_prob[:,   :,  i,  i:i  +
span].cpu().detach().numpy()
```

3. Model integration results

(1) Entity extraction

We found that when the maximum length (max_length) of the input text is different, the fusion of the resulting training models can achieve significant benefits. Additionally, we used a faster training version, Efficient GlobalPointer, to increase the diversity in the model fusion process. Using different backbones, the highest cross-validation score for 4-folds was 0.557, and the multi-model fusion score was 0.570, showing a significant improvement compared to a single model. The scores for single model and fusion fold-out predictions for entity extraction are shown in Table 5.1.

(2) Slot extraction

During the training process, using different backbones, the highest cross-validation score for 4-folds was 0.607, and the multi-model fusion score was 0.616, showing a significant improvement compared to a single model. Detailed results are shown in Table 5.2.

Table 5.1 Entity extraction single model and fusion fold-out prediction scores

Backbone	Head	max_length	4-fold fold-out prediction score	Fusion fold-out prediction score
Roformer	Efficient GlobalPointer	384	0.557	0.570
DeBERTa	GlobalPointer	280	0.556	
NEZHA	GlobalPointer	256	0.547	
RoBERTa	GlobalPointer	256	0.547	
MacBERT	GlobalPointer	256	0.549	

Table 5.2 Slot extraction single model and fusion fold-out prediction scores

Backbone	Head	max_length	4-fold fold-out prediction score	Fusion fold-out prediction score
Roformer	GlobalPointer	256	0.607	0.616
NEZHA	GlobalPointer	256	0.605	
RoBERTa	GlobalPointer	256	0.600	
MacBERT	GlobalPointer	256	0.602	

Reference

1. Su, J., Murtadha, A., Pan, S., Hou, J., Sun, J., Huang, W., Wen, B., & Liu, Y. (2022). *Global pointer: Novel efficient span-based approach for named entity recognition.* arXiv. https://papers.cool/arxiv/2208.03054

Chapter 6
Computer Vision (Image): Theoretical Part

The main purpose of computer vision is to enable computers to understand the content in image data, that is, to use cameras and computers to replace the human eye in recognizing, tracking, and measuring targets, among other tasks, and further perform image processing. This chapter mainly focuses on static image tasks, while video tasks will be discussed in Chap. 8. The input form of such tasks is three-dimensional image data (image height, image width, number of channels), and the common task types are as follows.

(1) Classification, which assigns a unique label to the input image.
(2) Semantic segmentation, which determines the classification of each pixel in the image.
(3) Object detection, which detects specific targets in the input image and frames the location of the target in the form of a rectangular box.
(4) Instance segmentation, which goes a step further than object detection, detecting targets in the input image and identifying the specific pixels of each target.

Generally speaking, image tasks based on deep learning can be processed using the general process shown in Fig. 6.1. This chapter firstly introduces each part of this process, then introduces some general techniques, and finally expands on different tasks, introducing commonly used models, loss functions, and various techniques.

6.1 General Process

6.1.1 Data Preprocessing

Data preprocessing refers to processing data according to certain rules to make it easier for subsequent analysis and computation. Preprocessing is divided into offline preprocessing and online preprocessing.

© Tsinghua University Press 2026
K. Xu, *Data Mining Competition Practices*,
https://doi.org/10.1007/978-981-95-3446-3_6

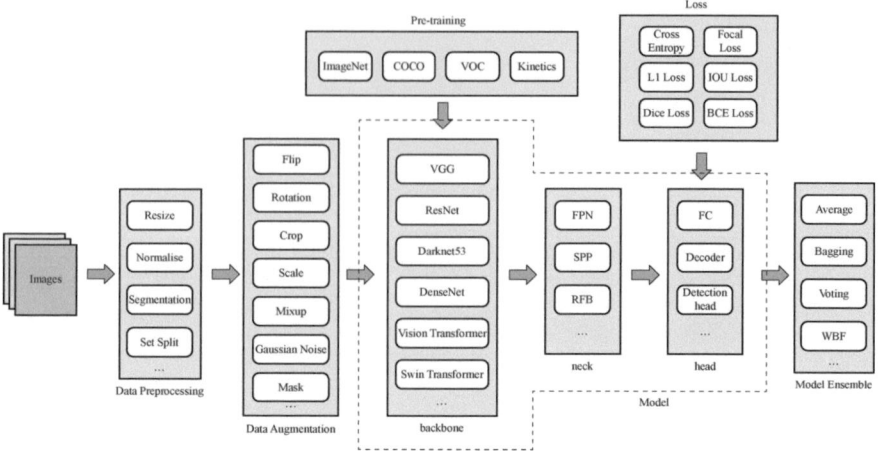

Fig. 6.1 General process of vision tasks

Offline preprocessing refers to manually processing data in advance. When the dataset is not good enough, human intervention in the data makes it easier for the model to learn. This includes data cleaning specified by manual rules, such as using tools like Cleanlab, algorithms, or manually cleaning noisy samples (noisy labels), merging categories, deleting similar or duplicate samples, deleting abnormal samples, performing ROI Crop (cutting irrelevant areas in large images), etc.; for imbalanced data, oversampling or undersampling is used to adjust the distribution; offline image enhancement, such as intensity normalization, high-pass filtering, color enhancement, etc.

Online preprocessing refers to processing the loaded images in advance during training and inference to make them conform to the network's input format or to speed up convergence.

The following code provides a simple example of image online preprocessing based on torchvision.

```
import torchvision.transforms as transforms
from PIL import Image
norm_mean = [0.485, 0.456, 0.406]
norm_std = [0.229, 0.224, 0.225]
# ImageNet mean & std
   transform = transforms.Compose([
   transforms.Resize(256),                # Resize the image
   transforms.CenterCrop(224),            # Center crop
   transforms.RandomFlip()                # Data augmentation
with random flip, detailed in Sect. 6.1.2
transforms.ToTensor(),                    # Convert to tensor
transforms.Normalize(norm_mean, norm_std), # Normalize
])
```

```
img_rgb = Image.open(img_path).convert('RGB')
img_t = transform(img_rgb)
```

6.1.2   Data Augmentation

Data augmentation is a technique that artificially expands the training dataset by performing transformations and combinations on the original data to generate more data. Deep learning models generally require sufficient training data for training, and when data is limited, data augmentation strategies can greatly expand the data volume, often significantly improving training effectiveness in practice.

Below are some common data augmentation methods and corresponding code.

Figure 6.2 shows the original image, and below it are the images obtained after a series of data augmentations to illustrate the principles of different data augmentation methods.

1. Flip

Apply random horizontal/vertical flip to the image, the code is as follows.

```
from torchvision import transforms
hflip = transforms.RandomHorizontalFlip(p=0.5)
vflip = transforms.RandomVerticalFlip(p=0.5)
```

Fig. 6.2 Original image

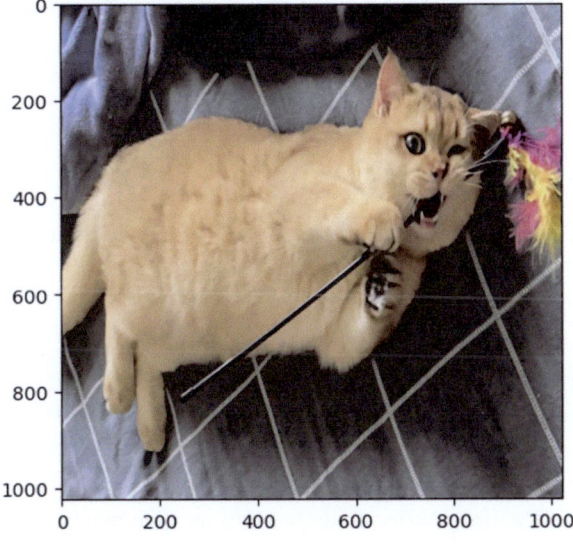

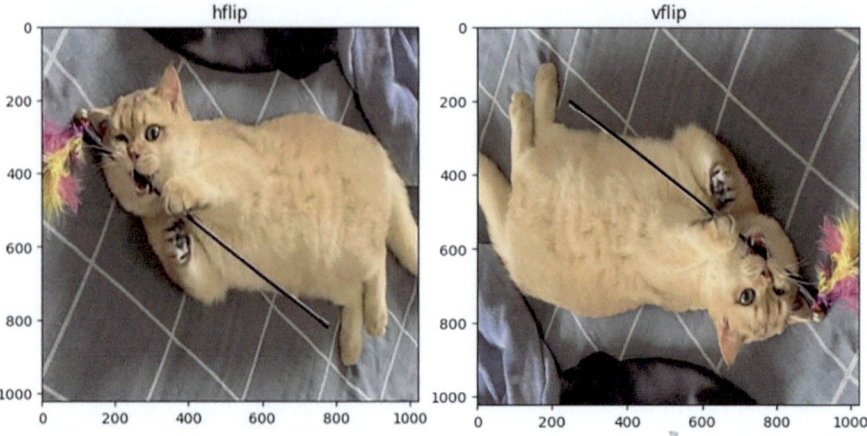

Fig. 6.3 Image after random flip

```
img_hflip = hflip(img)
img_vflip = vflip(img)
```

The enhanced images are shown in Fig. 6.3.

2. Rotation

Apply random angle rotation to the image, the code is as follows.

```
from torchvision import transforms
rot = transforms.RandomHorizontalFlip(degrees=(-180,180))
img_rot = rot(img)
```

The enhanced image is shown in Fig. 6.4.

3. Crop

Perform random cropping on the image, which may affect the semantic information of the image itself. Note that for detection and segmentation tasks, the label should be cropped in the same way. The code is as follows.

```
from torchvision import transforms
crop = transforms.RandomCrop(size=(224, 224))
img_crop = crop(img)
```

The image after random cropping is shown in Fig. 6.5.

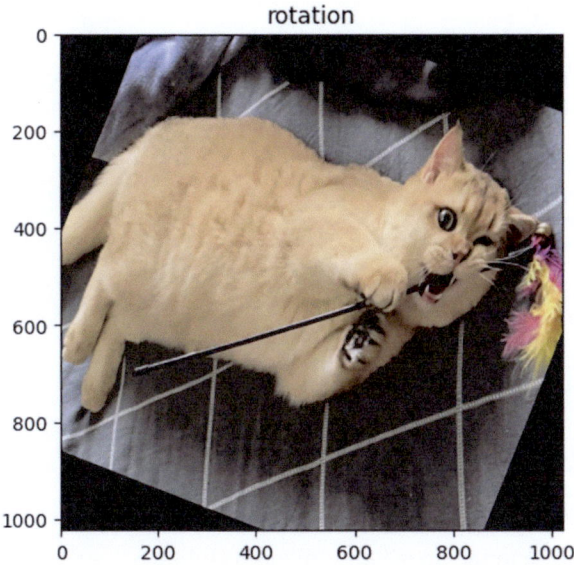

Fig. 6.4 Enhanced image

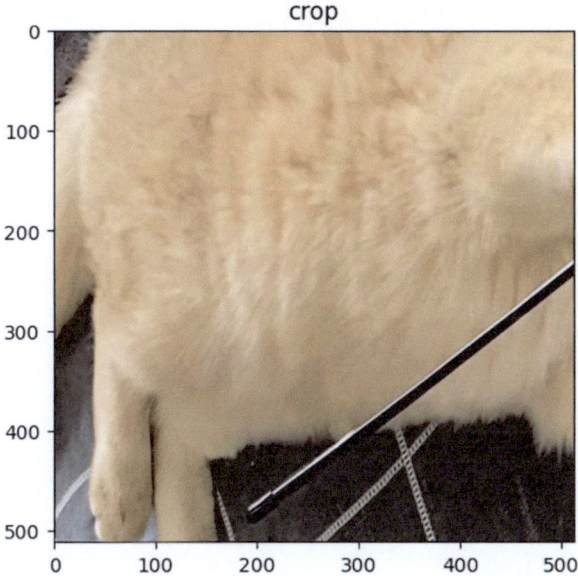

Fig. 6.5 Image after random cropping

4. Mixup

Overlay two different images with weighted summation, and overlay the labels with the same weights. The code is as follows.

```
def mixup_data(x, y, lam=1.):
    mixed_x = lam * x[0] + (1 - lam) * x[1]
    mixed_y = lam * y[0] + (1 - lam) * y[1]
        return mixed_x, mixed_y
img_mixup, label_mixup = mixup_data([img1, img2], [y1, y2])
```

The image after weighted summation is shown in Fig. 6.6.

Note: The mixup example here is a mixing method for classification. In object detection tasks and semantic segmentation tasks, mixup is also applicable.

5. Gaussian noise

Add Gaussian noise to the image, the code is as follows.

```
class GaussianNoise(object):
    def __init__(self,
            mean=0.0,
            std=1.0,
            amplitude=0.2,
            p=1):
        self.mean = mean
        self.std = std
        self.amplitude = amplitude
        self.p=p
    def __call__(self, img):
        if torch.rand(1).item() < self.p:
            h, w, c = img.shape
            N = self.amplitude * torch.normal(mean=self.mean,
                    std=self.std, size=(h, w, c))
            img = N + img
            img[img > 255] = 255
            return img
        else:
                return img
gaussian_noise = GaussianNoise()
img_gn = gaussian_noise(img)
```

The image after adding Gaussian noise is shown in Fig. 6.7.

6. ColorJitter

Randomly change the properties of the image, such as brightness, contrast, saturation, and hue. The code is as follows.

Fig. 6.6 Image after weighted summation

Fig. 6.7 Image after adding
Gaussian noise

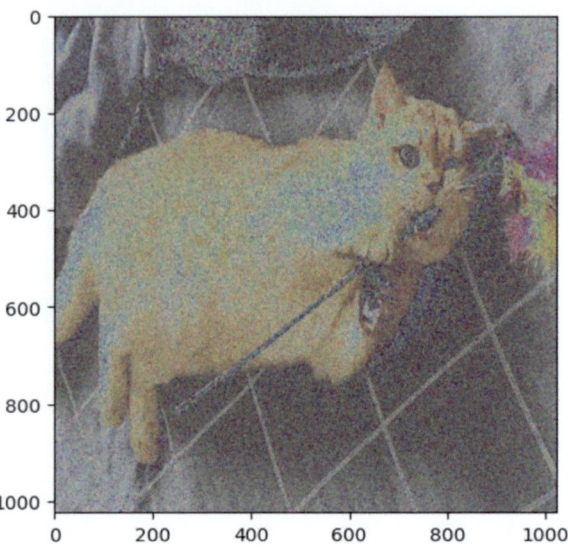

```
from torchvision import transforms
color_jitter   =   transforms.ColorJitter(brightness=0.5,
contrast=0.5, saturation=0.5, hue=0.5)
img_cj = color_jitter(img)
```

The image after ColorJitter is shown in Fig. 6.8.

Fig. 6.8 Image after
ColorJitter

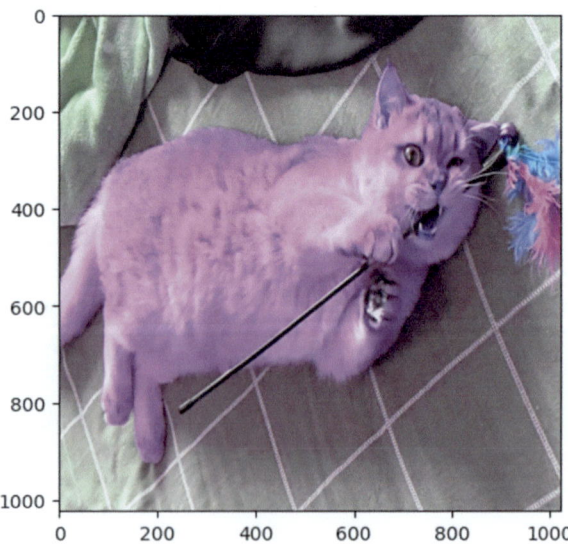

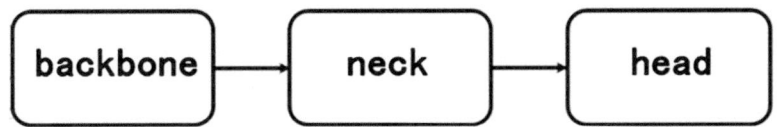

Fig. 6.9 Backbone-neck-head structure

In addition, methods such as Gaussian blur, salt-and-pepper noise, CutMix, and Mosaic can also be used for data augmentation, which will not be detailed here.

6.1.3 Pre-training

Pre-training refers to first training the model on other larger datasets, and then transferring the model parameters to downstream tasks for fine-tuning.

Models pre-trained on large datasets have already learned general image features, such as edges, textures, and shapes, which can help accelerate the convergence speed of the network and enhance generalization ability, allowing high-quality models to be trained even on small downstream datasets.

Depending on the task, different pre-training datasets are used, with ImageNet being the most commonly used visual pre-training dataset. Sometimes, self-supervised or unsupervised pre-training tasks can be designed to improve the model's representation capability.

ImageNet is one of the largest and most widely used open-source datasets in the CV field. It contains over 14 million manually annotated images. Generally, almost all types of tasks can use the ImageNet dataset for pre-training to improve performance.[1]

6.1.4 Model

Deep models in the CV field can generally be divided into a backbone-neck-head structure, as shown in Fig. 6.9.

The backbone-neck-head is a common architecture for deep models in the field of computer vision. The backbone part is the main body of the model, responsible for extracting the low-level features of the image. The neck part is responsible for further converting the features extracted by the backbone into higher-level features. The head part is the last layer of the model, responsible for converting the features extracted by the neck into the final output of the model.

---

[1] For a detailed introduction to ImageNet, see https://www.image-net.org/.

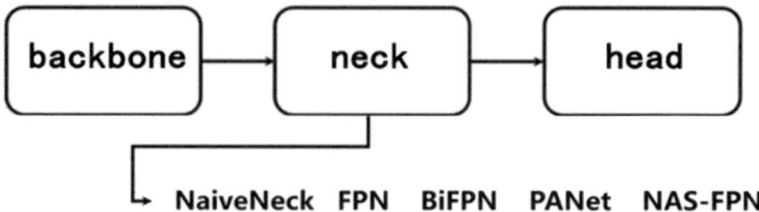

Fig. 6.10 Model structure and common neck schematic

1. Backbone

The backbone is the main network, generally serving as the feature extraction network in the model, and is the most important part of the network. It converts the input image data into high-level features, on which a series of subsequent tasks are performed, so the quality of feature representation directly determines the model's performance. In practice, the structure of the backbone is usually not designed independently, but rather mature structures that have been verified through extensive experiments are used. Common backbones include convolutional networks like ResNet, VGG, DenseNet, and transformer-based networks like ViT, Swin Transformer,[2] etc., and classic networks like ResNet are also integrated in torchvision.models.

Additionally, PyTorch implementations of most mainstream backbones can be found.

2. Neck

As shown in Fig. 6.10, the neck is the structure located between the backbone and the head, generally used to enhance the quality of feature extraction by the backbone. Common examples include Feature Pyramid Network (FPN), Spatial Pyramid Network (SPP), which are plug-and-play modules and sometimes not used. An appropriate neck module can significantly improve the model's performance.

3. Head

The head is the prediction head, which makes predictions using the features from the previous network. The classification head is generally highly related to the task type and loss function.

6.1.5 Loss Function

The loss function is the optimization target of the model, which is a function that quantifies the inconsistency between the model's predicted results and the true labels.

---

[2] The relevant source code can be found at https://github.com/huggingface/pytorch-image-models.

The loss function is highly related to the task form, and we will introduce several specific loss functions in detail in Sects. 6.2 to 6.4.

6.1.6 Ensemble Learning

In practice, we can train multiple different models and improve the final result by integrating the results of multiple models. Ensemble learning often achieves further improvements on the basis of the final result.

For classification tasks and semantic segmentation tasks, the averaging method is the most commonly used and simplest integration method. It can directly sum and average the class probabilities output by the models, or multiply by different weights and then average. Adjusting the weights of different models and monitoring performance on the validation set to select the optimal weights can often further enhance the integration effect. In addition, more complex methods such as Stacking can also be used for integration.

For object detection tasks, the averaging method is generally not applicable. Common integration methods for such tasks include non-maximum suppression (NMS), Soft-NMS, weighted boxes fusion (WBF), etc. We will introduce the specific process of WBF in Sect. 6.4.3.

6.1.7 General Techniques

1. Test-time data augmentation

As described in Sect. 6.1.2, various data augmentation techniques can not only be used to enhance the performance of the model during training, but also can be applied during testing to improve the performance. This technique is called Test Time Augmentation (TTA). The specific approach is to generate several copies of the image to be tested and apply data augmentation, typically rotation and flipping, and then input them into the model for inference separately. Integrate the inference results of multiple copies through methods such as averaging.

Essentially, TTA is also an ensemble strategy, which means integrating the different results obtained from different copies of the data using a single model without training more models.

An example of the pseudo code of TTA is as follows.

```
# img: The image to be tested
# model: The trained model
# final_pred: The final prediction of the model for img after
TTA
```

```
img1 = img
img2 = RandomFlip(img)
img3 = RandomRotation(img)
pred1, pred2, pred3 = model(img1), model(img2), model(img3)
final_pred = Ensemble(pred1, pred2, pred3)
```

2. SWA

Stochastic Weight Averaging (SWA) is to average the weights traversed by stochastic gradient descent (SGD) or any stochastic optimizer using a modified learning rate strategy, so as to obtain better convergence results, as shown in Fig. 6.11 [1]. Stochastic gradient descent tends to converge to a place where the loss is relatively low locally, but it is difficult to converge to the global minimum. If SWA is used, multiple weights can be weighted and averaged, thus it is possible to converge to a loss smaller than that of SGD.

Specifically, in the last few epochs of training, a relatively small and constant learning rate is used for training. Finally, the model weights of the last few epochs are averaged to obtain the final model. The use of SWA is already supported in PyTorch. The following is a simple example of using SWA in PyTorch.

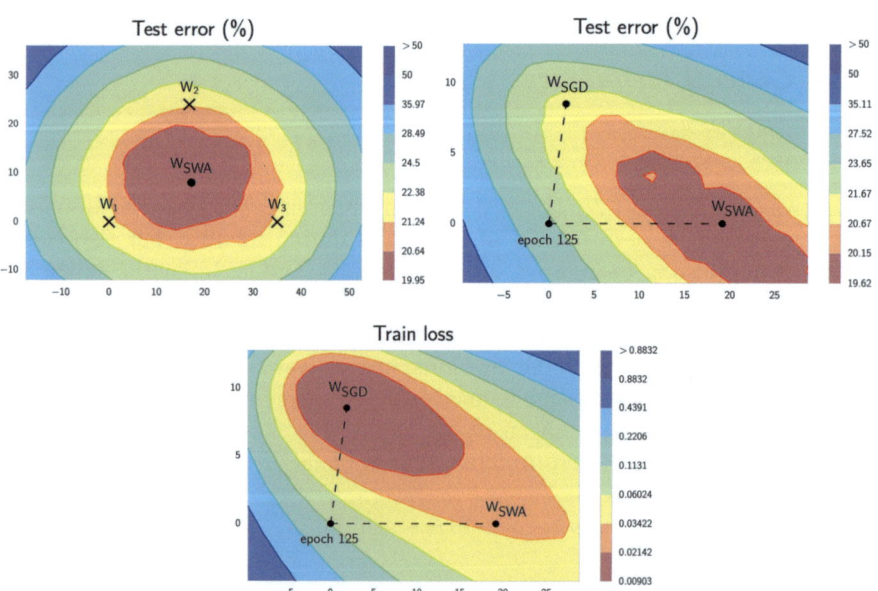

Fig. 6.11 Schematic diagram of the principle of SWA

```
from torch.optim.swa_utils import AveragedModel, SWALR
from torch.optim.lr_scheduler import CosineAnnealingLR
loader, optimizer, model, loss_fn =...
swa_model = AveragedModel(model)
scheduler = CosineAnnealingLR(optimizer, T_max=100)
swa_start = 5
swa_scheduler = SWALR(optimizer, swa_lr=0.05)
for epoch in range(100):
    for input, target in loader:
        optimizer.zero_grad()
        loss_fn(model(input), target).backward()
        optimizer.step()
    if epoch > swa_start:
        swa_model.update_parameters(model)
        swa_scheduler.step()
    else:
        scheduler.step()
torch.optim.swa_utils.update_bn(loader, swa_model)
preds = swa_model(test_input)
```

3. Knowledge distillation

In the knowledge distillation method, a teacher model is used to assist the training of the current model (e.g., the student model). The teacher model is a pre-trained model with high accuracy. Therefore, the student model can improve its accuracy while keeping the model complexity unchanged, which is very useful in scenarios with limited computational overhead. For example, ResNet-152 can be used as the teacher model to help train the student model ResNet-50. During the training process, a distillation loss is added to penalize the difference between the outputs of the student model and the teacher model.

Given an input, let p be the true probability distribution, and z and r be the features output by the student model and the teacher model respectively, which are the outputs of the last fully connected layer. Previously, we used the cross-entropy loss $\ell(p, \text{softmax}(z))$ to measure the difference between p and z. Here, the distillation loss also uses cross-entropy. So, the total loss function when using the knowledge distillation method is

$$\ell(p, \text{softmax}(z)) + T^2 \ell(\text{softmax}(r/T), \text{softmax}(z/T)).$$

In the above formula, the first term is the original loss function, and the second term is the distillation loss added to penalize the difference between the outputs of the student model and the teacher model. Here, T is a temperature hyperparameter used to make the output of softmax smoother. Experiments have shown that using ResNet-152 as the teacher model to train ResNet-50 can improve the accuracy of the latter.

The sample code is as follows.

```
# model1 is the student model and model2 is the teacher model
kl_loss = nn.KLDivLoss(reduction="batchmean")
ce_loss = nn.CrossEntropy()
for data, label in batches:
    output1 = model1(data)
    output2 = model2(data)
    loss1 = ce_loss(label, output1)
            loss2    =   kl_loss(F.log_softmax(output1,dim=1),
F.softmax(output2,dim=1))
    loss = loss1 + alpha * loss2
    ...
```

4. Pseudo-labels

After the model training is completed, predictions can be made on the test images or the unlabeled images provided in the competition (if any) to generate pseudo-labels. Mixing the data with pseudo-labels to the original training data and retraining the model usually leads to an improvement in the model performance. You can also use the new model to re-predict the unlabeled images, and keep iterating the previous steps until the model performance converges.

The sample process is as follows:

(1) Train the model using labeled data.
(2) Use the model to make predictions on unlabeled data. At this stage, TTA or ensemble learning techniques can be used.
(3) Take the predicted labels as pseudo labels.
(4) Mix the unlabeled data with the labeled data to obtain a new dataset.
(5) Retrain the model using the new dataset.

5. Learning Rate Scheduler

Training with a constant learning rate is usually not a good approach. Generally, some strategies need to be specified to dynamically adjust the learning rate during the training process. Commonly used methods include warmup, cosine annealing, etc., and the relevant learning rate adjustment modules can be found in torch.optim.lr_scheduler.

The sample code is as follows (cosine annealing + warmup):

```
import torch
import torch.optim as optim
import timm
# Define the model, optimizer and loss function
model = MyModel()
optimizer = optim.SGD(model.parameters(), lr=0.1)
```

```
loss_fn = nn.CrossEntropyLoss()
# Define the learning rate and learning rate scheduler
scheduler = timm.scheduler.CosineLRScheduler(optimizer =
optimizer,
              t_initial=200,
              # Total number of epochs for training
              lr_min=1e-5,
                  # Minimum value of learning rate decay for cosine
annealing
              warmup_t=4,
              # Number of epochs in the warmup stage
              warmup_lr_init=1e-4)
              # Initial value in the warmup stage
# Training loop
for epoch in range(100):
    # Adjust the learning rate
    scheduler.step(epoch)
    # Train the model
    train_loss, train_acc = train(model, optimizer, loss_fn,
train_data)
    # Validate the model
    val_loss, val_acc = validate(model, loss_fn, val_data)
    # Print the metrics
    print(f'Epoch {epoch}: train loss = {train_loss:.3f}, train
acc = {train_acc:.3f}, val loss = {val_loss:.3f}, val acc =
{val_acc:.3f}')
```

6. Others

The following lists some other practical experiences for reference:

- Ensure that the data distribution and label distribution of the training set and the test set are consistent.
- Pay attention to whether the classes are balanced. If not, class-specific sampling or weighted loss should be carried out.
- Larger geometric transformations can be tried, such as elastic transformation, affine transformation, spline affine transformation, and Pillow distortion.
- Apply channel shuffling, that is, randomly shuffle the channel order.
- Try different learning rates.
- Try different batch sizes.
- Excessive augmentation will reduce the accuracy.
- Divide the training set and the test set unevenly according to classes.

6.2 Classification Tasks

6.2.1 Task Introduction and Common Models

Image classification is one of the most fundamental tasks in the field of computer vision, with the main goal of automatically categorizing an image into a specific category. For example, classifying a photo as a category of dog, cat, or other animals, or classifying a street scene photo as a park, store, or residential area, etc. The most important metric for classification tasks is accuracy, which is the proportion of correctly classified images by the model to the total number.

The commonly used baseline models for image classification are as follows.

1. ResNet

Residual network (ResNet) is a type of deep convolutional neural network proposed by the Microsoft Research Asia team in 2015. In ResNet, the output of each convolutional layer is added to the identity mapping of the input, which can avoid the problem of gradient vanishing or gradient explosion in deep networks, allowing deeper networks to be trained. An important component of ResNet is the residual block, which contains two convolutional layers and an identity mapping. The residual block passes the input of the previous layer directly to the subsequent layers through cross-layer connections, thereby increasing the depth of the network. This is the most classic network structure and is usually used as a baseline model.

2. ViT

Vision transformer (ViT) is a model that uses the transformer structure for image classification, proposed by the Google team. ViT divides the input image into several small image patches and unfolds them into a one-dimensional sequence, then processes them through the transformer model, and finally outputs the category of the image. By introducing the self-attention mechanism, ViT allows the model to automatically focus on image features relevant to the current task, thereby improving the model's accuracy. ViT performs well in many vision tasks, such as image classification, object detection, and so on.

3. Swin Transformer

Swin Transformer is an image classification model based on the transformer structure, proposed by researchers from the Chinese University of Hong Kong and Huawei. Swin Transformer adopts a hierarchical attention mechanism, dividing the input image into multiple layers, where regions in each layer are interrelated with other images in the same layer, thereby improving the model's accuracy. Additionally, Swin Transformer employs a local window mechanism and cross-layer path mechanism, enabling the model to better capture local and global features in images.

6.2.2 Loss Functions

Commonly used loss functions for classification tasks include cross-entropy and focal loss.

1. Cross-entropy

Cross-entropy is one of the most common classification loss functions, used to evaluate the gap between predicted categories and true categories in classification tasks. The cross-entropy loss function can be expressed as

$$J_{CE} = -\sum_{i=1}^{N}(y_i \log(\hat{y}_i) + (1 - y_i) \log(1 - \hat{y}_i)),$$

where y_i is the true label, $\hat{y}_i$ is the probability predicted by the model.

The PyTorch implementation code for cross-entropy is as follows.

```
from torch import nn
criterion= nn.CrossEntropyLoss()
loss = criterion(y_pred, y_true)
```

2. Focal loss

Focal loss, proposed by Kaiming He, is a loss function for scenarios with class imbalance. Focal loss addresses the model training issues caused by uneven sample distribution from the perspective of sample classification difficulty. The specific mathematical form of focal loss is

$$J_{FL} = -(1 - \hat{y}_i)^\gamma \log \hat{y}_i$$

where $\hat{y}_i$ is the probability predicted by the model.

The PyTorch implementation (example) code for focal loss is as follows.

```
class WeightedFocalLoss(nn.Module):
    def __init__(self, alpha=.25, gamma=2):
        super(WeightedFocalLoss, self).__init__()
        self.alpha = torch.tensor([alpha, 1-alpha]).cuda()
        self.gamma = gamma
    def forward(self, inputs, targets):
        BCE_loss = F.binary_cross_entropy_with_
logits(inputs,
targets, reduction='none')
        targets = targets.type(torch.long)
        at = self.alpha.gather(0, targets.data.view(-1))
```

```
pt = torch.exp(-BCE_loss)
F_loss = at*(1-pt)**self.gamma * BCE_loss
return F_loss.mean()
```

6.2.3 Common Techniques

1. Label smoothing

Label smoothing is a regularization method used during training, typically in classification problems. The main purpose of label smoothing is to prevent the model from being overly confident in predicting labels during training, thereby improving the issue of poor generalization ability. Label smoothing involves adding a smoothing coefficient to the one-hot labels, i.e.,

$$q_i = \begin{cases} 1 - \varepsilon, & i = y \\ \varepsilon/(K - 1), & i \neq y \end{cases},$$

where q_i is the smoothed label, and y is the original label.

The example code is as follows.

```
# Get the model's prediction
pred = model(inputs)
# Get the model's prediction
pred = model(inputs)
def smooth_one_hot(true_labels: torch.Tensor, classes: int,
smoothing=0.0):
    """
    true_labels are the original one-hot labels, classes is
the total number of categories, and smoothing is the smoothing
factor.
    """
    assert 0 <= smoothing < 1
    confidence = 1.0 - smoothing
    label_shape = torch.Size((true_labels.size(0), classes))
    with torch.no_grad():
        true_dist = torch.empty(size=label_shape, device=true_
labels.device)
        true_dist.fill_(smoothing / (classes - 1))
        _, index = torch.max(true_labels, 1)
        true_dist.scatter_(1, torch.LongTensor(index.unsqueeze(1)),
confidence)
    return true_dist
```

```
# Calculate the loss using smoothed labels
smooth_loss = loss_fn(smooth_one_hot(labels), predictions)
```

2. Weighted combination of multiple losses

Using a weighted combination of focal loss and cross-entropy loss as the optimization target can sometimes improve the model's robustness and achieve better
generalization performance.

The example code is shown below.

```
# Calculate the weighted sum of Focal Loss and cross-entropy
loss
# Within the training loop
    ...
    logits = model(data)
    f_loss = focal_loss(logits, labels)
    ce_loss = cross_entropy_loss(logits, labels)
    total_loss = ce_loss * ce_weight + f_loss * focal_weight
    total_loss.backward()
    ...
```

6.3 Segmentation Tasks

6.3.1 Task Introduction and Common Models

Semantic segmentation tasks require assigning a category label to each pixel in
an image, which can be understood as a pixel-by-pixel classification task. Unlike
image classification tasks, semantic segmentation tasks need to output a category
map of the same size as the original image, where the value of each pixel represents
the category of the corresponding position. The model needs to learn a pixel-level
feature representation to assign each pixel in the image to the correct category.

Common baseline models for semantic segmentation are as follows.

1. U-Net

U-Net is a semantic segmentation model based on convolutional neural networks,
proposed by Olaf Ronneberger et al. in 2015. U-Net adopts a network architecture
called a U-shaped structure, which includes a downsampling path and an upsampling path. In the downsampling path, the network extracts features through pooling
and convolution operations; in the upsampling path, the network restores features
to the original size through deconvolution and skip connection operations, thereby

achieving pixel-level segmentation. U-Net has been widely used in fields such as medical image segmentation.

2. U-Net++

U-Net++ is an improved semantic segmentation model based on U-Net, which has been widely used in recent semantic segmentation competitions due to its improved segmentation accuracy. U-Net++ adopts a multi-resolution branch and dense skip connection mechanism to further improve the model's segmentation accuracy. Specifically, the network architecture of U-Net++ consists of several U-Net modules, each of which is composed of an encoder and a decoder, similar to the U-shaped structure of U-Net. In U-Net++, the feature maps of the encoder part are not only directly passed to the decoder but also skip-connected to feature maps of other resolutions within the same module, thereby enhancing the model's feature representation capability. Additionally, a dense skip connection mechanism is introduced in the decoder of each U-Net module, allowing the model to better learn local features.

3. DeepLab

DeepLab is a semantic segmentation model proposed by the Google team, which employs techniques such as atrous convolution and multi-scale pooling to enhance the model's receptive field and segmentation accuracy. Atrous convolution can increase the receptive field by introducing holes into the convolution kernel, thereby effectively capturing global features in the image. Multi-scale pooling can improve the model's ability to recognize objects of different scales by pooling feature maps of different scales. DeepLab has evolved to the third generation (DeepLabV3+) and has performed excellently in many semantic segmentation tasks. Additionally, models like ViT and SwinTransformer can also be used for semantic segmentation tasks by modifying the model configuration.

6.3.2 Loss Function

Semantic segmentation can use the same loss functions as classification tasks, such as pixel-wise cross-entropy. Besides, there are loss functions that directly optimize the metrics of semantic segmentation tasks (IoU), such as dice loss. Dice loss is named after the dice coefficient, which is a metric function used to evaluate the similarity between two samples. The larger the value, the more similar the two samples are. The mathematical expression of the dice coefficient is as follows.

$$s = \frac{|X \cap Y|}{|X| + |Y|}.$$

Dice loss can be expressed as

$$J_{\text{Dice}} = 1 - \frac{2|X \cap Y|}{|X| + |Y|}.$$

In the formula, X represents the pixel labels of the true segmentation image, Y represents the pixel categories of the model-predicted segmentation image, $|X \cap Y|$ is the dot product between the pixels of the predicted image and the true label image, with the dot product results summed, and $|X|$ and $|Y|$ are the sums of the corresponding pixels in their respective images.

The PyTorch code implementation of dice loss is as follows.

```
import torch.nn as nn
import torch.nn.functional as F
class DiceLoss(nn.Module):
    def __init__(self, weight=None, size_average=True):
        super(DiceLoss, self).__init__()
    def forward(self, logits, targets):
        num = targets.size(0)
        smooth = 1
        probs = F.sigmoid(logits)
        m1 = probs.view(num, -1)
        m2 = targets.view(num, -1)
        intersection = (m1 * m2)
        score = 2. * (intersection.sum(1) + smooth) / (m1.sum(1)
+ m2.sum(1) + smooth)
        score = 1 - score.sum() / num
        return score
```

6.3.3 Common Techniques

1. Apply morphological algorithms for post-processing (opening/closing operations)

Opening operations can eliminate smaller isolated objects and smooth edges; closing operations can eliminate small holes. The effectiveness of morphological operations is related to model performance, tasks, kernel size, etc. Improper use may degrade segmentation results. It is recommended to visualize segmentation results before applying these operations, analyze in combination with output conditions and task characteristics, and experiment on the validation set first.

Example code is shown below.

```
# Model prediction
mask = model(input)
# Apply morphological algorithms for post - processing
# Define the kernel size. The larger the kernel, the stronger
the effect of the morphological operation.
kernel = np.ones((3, 3), np.uint8)
# Opening operation
opening = cv2.morphologyEx(mask, cv2.MORPH_OPEN, kernel,
iterations=2)
# Closing operation
closing = cv2.morphologyEx(mask, cv2.MORPH_CLOSE, kernel,
iterations=2)
```

2. Sliding window partitioning for high-resolution images

For high-resolution images, you can first use a sliding window to partition the image into a set of smaller images, perform inference on each, and then aggregate the inference results of all images together. Since partitioning may split some objects into two different image blocks, there should be overlap between the two windows. Example code is shown below.

```
# Define the sliding window size and stride. The stride is
smaller than the window size to make the windows overlap.
stride = 768
win_size = 1024
total_col = max(int((img.shape[1] - win_size + stride - 1) /
stride) + 1, 1)
total_row = max(int((img.shape[0] - win_size + stride - 1) /
stride) + 1, 1)
# Create an empty inference result
total_preds = np.zeros(shape=(num_class, img.shape[0],
img.shape[1]), dtype=np.int32)
# Record which regions have been calculated multiple times
count = np.zeros(shape=(num_class, img.shape[0],
img.shape[1]), dtype=np.int32)
for row in range(total_row):
   for col in range(total_col):
       patch = img[row * stride:row * stride + win_size, col *
stride: col * stride + win_size, :]
       preds = model(patch)
       total_preds[row * stride:row * stride + win_size, col *
stride:col * stride + win_size, :] += preds
      count[row * stride:row * stride + win_size, col * stride:col
* stride + win_size, :] += 1
```

```
# Take the average of multiple calculations for the overlapping
parts
total_preds /= count
```

3. Combined loss function

As mentioned in Sect. 6.2.3, using multiple loss functions can improve the robustness of the model, which is also applicable in segmentation tasks, such as using the weighted sum of BCE+Dice+Focal loss as the loss function.

The sample code is as follows.

```
# in train loop
  ...
  preds = model(data)
  dice_loss = dice_loss(preds, labels)
  ce_loss = cross_entropy_loss(preds, labels)
  f_loss = focal_loss(preds, labels)
  total_loss = ce_loss * ce_weight + f_loss * focal_weight +
dice_loss *
dice_weight
  total_loss.backward()
  ...
```

6.4 Detection Tasks

6.4.1 Task Introduction and Common Models

Object detection requires detecting the position, size, and category of objects in images or videos. Unlike image classification and semantic segmentation, object detection needs to output both the position and category information of the target and can detect multiple targets simultaneously.

The object detection task is usually divided into two stages: object extraction and object classification. Object extraction typically uses methods like sliding windows or anchor boxes to extract candidate objects at different positions and sizes. Object classification involves classifying the specific category of the candidate objects.

Common models for object detection include region-based models and single-stage detection models. Region-based models include R-CNN, Fast R-CNN, Faster R-CNN, Mask R-CNN, etc. These models usually adopt a two-stage process, first generating candidate objects through region extraction methods, and then classifying and locating them using deep learning models. These models perform well in terms of accuracy but are slower in computation.

Single-stage detection models directly detect and classify objects in images, with common models including YOLO, SSD, etc. These models typically use a method called "anchor boxes" to directly predict the position and category information of targets in the image. The YOLO series is currently one of the most widely used and effective object detection methods.

You Only Look Once (YOLO) series is a baseline model for common object detection tasks.

YOLO is a series of single-stage detection models that use techniques like "anchor boxes" and multi-scale feature maps to detect and classify objects in the entire image in a single forward pass, offering high detection speed and real-time performance, thus being widely applied in practice. The network architecture of the YOLO series models can be divided into two parts: feature extraction network and detection network. The feature extraction network uses convolutional neural networks to extract features from the input image and generate multi-scale feature maps. The detection network performs object detection and classification based on the multi-scale feature maps, usually implemented with convolutional layers and fully connected layers.

The characteristic of the YOLO series models is the use of the "anchor boxes" method, which pre-defines some fixed-size and ratio rectangular boxes to segment and classify objects in the input image. The YOLO series models also use the multi-scale feature map method, which can handle targets of different sizes and scales, thus achieving detection and classification of the entire image.

Currently, the YOLO series models have released multiple versions, including YOLO v1 to v7. Each version has been improved and optimized based on the original, thereby enhancing detection accuracy and speed. For example, YOLO v3 uses Darknet-53 as the feature extraction network and introduces technologies like FPN and Path Aggregation Network (PAN), further improving detection accuracy and speed.

6.4.2 Loss Function

As mentioned above, object detection is divided into detection and classification parts, requiring different loss functions for optimization, namely classification loss and position loss. Classification loss usually uses cross-entropy, focal loss, or a weighted combination of both, as detailed in Sect. 6.2 on classification tasks. Below is an introduction to common position losses.

1. L1 Loss

L1 loss (MAE) is the absolute value of the difference between the predicted value and the true value, i.e., the mean absolute error. The formula is

$$\text{MAE} = \frac{1}{n} \sum_{i=1}^{n} |\hat{y}_i - y_i|$$

```
from torch import nn
criterion = nn.L1Loss()
loss = criterion(y_pred, y_true)
```

2. L2 loss

L2 Loss (MSE) is the square of the difference between the predicted value and the true value. The formula is as follows:

$$\text{MSE} = \frac{1}{n} \sum_{i=1}^{n} (\hat{y}_i - y_i)^2$$

```
from torch import nn
criterion = nn.MSELoss()
loss = criterion(y_pred, y_true)
```

3. IoU loss

IoU loss (IoU loss) is calculated based on the intersection over union (IoU) of the predicted box and the ground truth box. Let the predicted box be P, and the ground truth box be G, then IoU can be expressed as follows:

$$\text{IoU} = \frac{P \cap G}{P \cup G}.$$

Then IoU loss can be expressed as follows:

$$\text{IoU Loss} = 1 - \text{IoU}.$$

The implementation code for IoU loss is as follows.

```
def IOU(box1, box2):
    """
    IoU Loss
        :param box1: tensor [batch, w, h, num_anchor, 4], xywh
    predicted values
```

```
    :param box2: tensor [batch, w, h, num_anchor, 4], xywh true
values
    :return: tensor [batch, w, h, num_anchor, 1]
    """
    box1_xy, box1_wh = box1[..., :2], box1[..., 2:4]
    box1_wh_half = box1_wh / 2.
    box1_mines = box1_xy - box1_wh_half
    box1_maxes = box1_xy + box1_wh_half
    box2_xy, box2_wh = box2[..., :2], box2[..., 2:4]
    box2_wh_half = box2_wh / 2.
    box2_mines = box2_xy - box2_wh_half
    box2_maxes = box2_xy + box2_wh_half
    # Calculate IoU for all true and predicted values
    intersect_mines = torch.max(box1_mines, box2_mines)
    intersect_maxes = torch.min(box1_maxes, box2_maxes)
    intersect_wh = torch.max(intersect_maxes-intersect_mines,
torch.zeros_like(intersect_maxes))
    intersect_area = intersect_wh[..., 0]*intersect_wh[..., 1]
    box1_area = box1_wh[..., 0]*box1_wh[..., 1]
    box2_area = box2_wh[..., 0]*box2_wh[..., 1]
    union_area = box1_area+box2_area-intersect_area
    iou = intersect_area / torch.clamp(union_area, min=1e-6)
    return iou
```

4. GIoU loss

When the predicted box and the ground truth box do not intersect, IoU is always 0, resulting in IoU loss being always 1, which prevents the network from training. GIoU loss (GIoU loss) adds a penalty term to the IoU loss, and its specific definition is as follows:

$$\text{GIoU Loss} = 1 - \text{IoU} + \frac{|C - P \cup G|}{|C|},$$

where C is the smallest enclosing box of P and G, that is, the smallest rectangle that can completely enclose P and G.

The implementation code for GIoU loss is as follows.

```
def GIOU(box1, box2):
    """
    GIoU Loss
    :param box1: tensor [batch, w, h, num_anchor, 4], xywh
predicted values
    :param box2: tensor [batch, w, h, num_anchor, 4], xywh true
values
    :return: tensor [batch, w, h, num_anchor, 1]
```

```
    """
            b1_x1,   b1_x2   =   box1[...,0]   -   box1[...,2]/
2,box1[...,0]+box1[...,2]/2
            b1_y1,   b1_y2   =   box1[...,1]   -   box1[...,3]/
2,box1[...,1]+box1[...,3]/2
            b2_x1,   b2_x2   =   box2[...,0]   -   box2[...,2]/
2,box2[...,0]+box2[...,2]/2
            b2_y1,   b2_y2   =   box2[...,1]   -   box2[...,3]/
2,box2[...,1]+box2[...,3]/2
    box1_xy, box1_wh = box1[..., :2], box1[..., 2:4]
    box1_wh_half = box1_wh / 2.
    box1_mines = box1_xy - box1_wh_half
    box1_maxes = box1_xy + box1_wh_half
    box2_xy, box2_wh = box2[..., :2], box2[..., 2:4]
    box2_wh_half = box2_wh / 2.
    box2_mines = box2_xy - box2_wh_half
    box2_maxes = box2_xy + box2_wh_half
    # Calculate IoU for all true and predicted values
    intersect_mines = torch.max(box1_mines, box2_mines)
    intersect_maxes = torch.min(box1_maxes, box2_maxes)
    intersect_wh = torch.max(intersect_maxes-intersect_mines,
torch.zeros_like(intersect_maxes))
    intersect_area = intersect_wh[..., 0]*intersect_wh[..., 1]
    box1_area = box1_wh[..., 0]*box1_wh[..., 1]
    box2_area = box2_wh[..., 0]*box2_wh[..., 1]
    union_area = box1_area+box2_area-intersect_area
    iou = intersect_area / torch.clamp(union_area, min=1e-6)
    # Calculate the width and height of the minimum
    cw = torch.max(b1_x2, b2_x2) - torch.min(b1_x1, b2_x1)
    ch = torch.max(b1_y2, b2_y2) - torch.min(b1_y1, b2_y1)
    c_area = cw * ch + 1e-16  # convex area
    return iou - (c_area - union_area) / c_area
```

5.   DIoU Loss

The (distance IoU loss (DIoU loss) takes into account the distance between the centers of the bounding boxes, and its specific definition is as follows.

$$\text{DIoU Loss} = 1 - \text{IoU} + \frac{\rho^2(P, G)}{c^2}.$$

Among them, c is the diagonal length of the minimum enclosing box of P and G, and ρ represents the distance between the centers of P and G.

The implementation code of the DIoU loss is as follows.

```
def DIOU(box1, box2):
    """
    DIoU loss
       :param box1: tensor [batch, w, h, num_anchor, 4], xywh
    predicted values
       :param box2: tensor [batch, w, h, num_anchor, 4], xywh
    true values
       :return: tensor [batch, w, h, num_anchor, 1]
    """
            b1_x1,   b1_x2   =   box1[...,0]   -   box1[...,2]/
    2,box1[...,0]+box1[...,2]/2
            b1_y1,   b1_y2   =   box1[...,1]   -   box1[...,3]/
    2,box1[...,1]+box1[...,3]/2
            b2_x1,   b2_x2   =   box2[...,0]   -   box2[...,2]/
    2,box2[...,0]+box2[...,2]/2
            b2_y1,   b2_y2   =   box2[...,1]   -   box2[...,3]/
    2,box2[...,1]+box2[...,3]/2
    box1_xy, box1_wh = box1[..., :2], box1[..., 2:4]
    box1_wh_half = box1_wh / 2.
    box1_mines = box1_xy - box1_wh_half
    box1_maxes = box1_xy + box1_wh_half
    box2_xy, box2_wh = box2[..., :2], box2[..., 2:4]
    box2_wh_half = box2_wh / 2.
    box2_mines = box2_xy - box2_wh_half
    box2_maxes = box2_xy + box2_wh_half
    # Calculate IoU for all true and predicted values
    intersect_mines = torch.max(box1_mines, box2_mines)
    intersect_maxes = torch.min(box1_maxes, box2_maxes)
    intersect_wh = torch.max(intersect_maxes-intersect_mines,
    torch.zeros_like(intersect_maxes))
    intersect_area = intersect_wh[..., 0]*intersect_wh[..., 1]
    box1_area = box1_wh[..., 0]*box1_wh[..., 1]
    box2_area = box2_wh[..., 0]*box2_wh[..., 1]
    union_area = box1_area+box2_area-intersect_area
    iou = intersect_area / torch.clamp(union_area, min=1e-6)
    # Calculate the width and height of the minimum enclosing box
    cw = torch.max(b1_x2, b2_x2) - torch.min(b1_x1, b2_x1)  #
    convex
    (smallest enclosing box) width
    ch = torch.max(b1_y2, b2_y2) - torch.min(b1_y1, b2_y1)
    c2 = cw ** 2 + ch ** 2 + 1e-16
    # Square of the distance between the centers of the two boxes
    rho2 = ((b2_x1 + b2_x2) - (b1_x1 + b1_x2)) ** 2 / 4 + ((b2_y1
    + b2_y2) - (b1_y1 + b1_y2)) ** 2 / 4
    return iou - rho2 / c2
```

6. CIoU loss

CIoU considers the following metrics on the basis of DIoU: overlap area, center point distance, and aspect ratio consistency, specifically defined as follows.

$$\text{CIoULoss} = 1 - \text{IoU} + \frac{\rho^2(P, G)}{c^2} + \alpha v$$

where:

$$v = \frac{4}{\pi^2} \left(\tan^{-1} \frac{\omega^g}{h^g} - \tan^{-1} \frac{\omega^p}{h^p} \right)^2$$

$$\alpha = \frac{v}{(1 - \text{IoU}) + v}.$$

The implementation code for CIoU loss is as follows.

```python
def CIOU(box1, box2):
    """
    CIoU loss
        :param box1: tensor [batch, w, h, num_anchor, 4], xywh
    predicted values
        :param box2: tensor [batch, w, h, num_anchor, 4], xywh ground
    truth values
        :return: tensor [batch, w, h, num_anchor, 1]
    """
    box1_xy, box1_wh = box1[..., :2], box1[..., 2:4]
    box1_wh_half = box1_wh / 2.
    box1_mines = box1_xy - box1_wh_half
    box1_maxes = box1_xy + box1_wh_half
    box2_xy, box2_wh = box2[..., :2], box2[..., 2:4]
    box2_wh_half = box2_wh / 2.
    box2_mines = box2_xy - box2_wh_half
    box2_maxes = box2_xy + box2_wh_half
    # Calculate IoU for all true and predicted values
    intersect_mines = torch.max(box1_mines, box2_mines)
    intersect_maxes = torch.min(box1_maxes, box2_maxes)
    intersect_wh = torch.max(intersect_maxes-intersect_mines,
torch.zeros_like(intersect_maxes))
    intersect_area = intersect_wh[..., 0]*intersect_wh[..., 1]
    box1_area = box1_wh[..., 0]*box1_wh[..., 1]
    box2_area = box2_wh[..., 0]*box2_wh[..., 1]
    union_area = box1_area+box2_area-intersect_area
    iou = intersect_area / torch.clamp(union_area, min=1e-6)
    # Calculate the difference of the centers
    center_distance = torch.sum(torch.pow((box1_xy-box2_xy),
2), dim=-1)
```

```
    # Find the top-left and bottom-right points of the minimum
enclosing box of the two boxes
      enclose_mines = torch.min(box1_mines, box2_mines)
      enclose_maxes = torch.max(box1_maxes, box2_maxes)
         enclose_wh = torch.max(enclose_maxes-enclose_mines,
torch.zeros_like(intersect_maxes))
      # Calculate the diagonal distance
       enclose_diagonal = torch.sum(torch.pow(enclose_wh, 2),
dim=-1)
      ciou = iou - 1. * center_distance / torch.clamp(enclose_
diagonal, min=1e-6)
      v = (4/(math.pi**2))*torch.pow((torch.atan(box1_wh[...,
0]/torch.clamp(box1_wh[..., 1], min=1e-6))-torch.atan(box2_
wh[..., 0]/torch.clamp(box2_wh[..., 1], min=1e-6))), 2)
      alpha = v / torch.clamp((1.-iou+v), min=1e-6)
      ciou = ciou - alpha * v
      return ciou
```

6.4.3 Common Techniques

1. Weighted box fusion

Compared to the most commonly used non-maximum suppression integration, weighted box fusion (WBF) is a better integration method, which performs clustering-like operations and weighted averaging on multiple bounding boxes (bbox) based on confidence and IoU. The specific process is as follows.

(1) Initialize the list B, add all the bounding boxes (bboxes) predicted by multiple models to the list B, and sort them in descending order of confidence.
(2) Declare empty lists L and F. Each element in L is a list, and each list contains the bboxes within a cluster. Each element in F represents the representative bbox of a cluster.
(3) Traverse the bboxes in B in descending order of confidence. For each bbox, use the rule that the intersection over union (IoU) is greater than a threshold for matching, and try to match this bbox with the bboxes in F. If a match is found, add this bbox to the cluster corresponding to the matched bbox in F. If no match is found, add this bbox as the representative of a new cluster to F, and also add a new cluster list to L.
(4) For each cluster in L, generate a new bbox. Calculate the confidence, size, and coordinates of the new bbox according to the following rules, where T is the number of bboxes within the cluster.

$$C = \frac{\sum_{i=1}^{T} C_i}{T}, X_{1,2} = \frac{\sum_{i=1}^{T} C_i \cdot X_{1,2_i}}{\sum_{i=1}^{T} C_i}, Y_{1,2} = \frac{\sum_{i=1}^{T} C_i \cdot Y_{1,2_i}}{\sum_{i=1}^{T} C_i}.$$

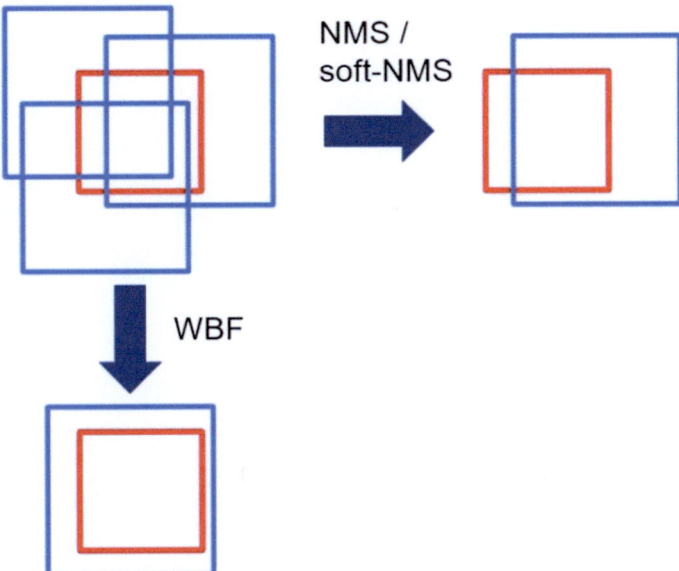

Fig. 6.12 Schematic diagram of the difference between WBF and NMS

(5) Scale the confidence of each new bbox according to the following rules, where
T is the number of bboxes in the cluster.

$$C = C \times \frac{\min(T, N)}{N}.$$

Figure 6.12 [2] shows the difference between weighted boxes fusion (WBF) and
non-maximum suppression (NMS) for box integration

To implement WBF, you can use the module provided by the paper's author and
install it via the "pip install ensemble-boxes" command. The sample code for usage
is as follows.

```
from ensemble_boxes import *
# Fuse the bounding boxes predicted by the same model
boxes, scores, labels = weighted_boxes_fusion(
                [boxes_list],
                [scores_list],
                [labels_list],
                weights=None,
                method=method,
                iou_thr=iou_thr,
                thresh=thresh)
```

2. Stuffing data augmentation

You can cut out the objects in one image and directly paste them into another image, and the effect is better when there are mask-level annotations. The sample code is as follows.

```
import os
from skimage import io
import numpy as np
import random
import tifffile as tif
import matplotlib.pyplot as plt
from tqdm import tqdm
import cv2
def duck_stuffing(target_img, target_label, available_img_
dir, available_label_dir):
    target_img = target_img.copy()
    target_label = target_label.copy()
    available_length = len(os.listdir(available_img_dir))
    num_ins = np.max(target_label)
    flag = True
    while flag or random.randint(0, 9) < 3:
        flag = False
        num_ins += 1
        # Randomly select another image
        choice = random.randint(0, available_length - 1)
        choose_img = os.listdir(available_img_dir)[choice]
        src_img = io.imread(os.path.join(available_img_dir,
choose_img))
        src_label = tif.imread(os.path.join(available_label_
dir, choose_img.replace('.jpg', '.tif')))
        # Randomly select an instance
        choice = random.randint(1, np.max(src_label))
        nonzero = (src_label == choice).nonzero()
        x_min = nonzero[0].min()
        y_min = nonzero[1].min()
        x_max = nonzero[0].max()
        y_max = nonzero[1].max()
        src_img = src_img[x_min:x_max, y_min:y_max]
        src_label = src_label[x_min:x_max, y_min:y_max]
        # Randomly transform the instance
        rot = random.randint(0, 3)
        flip = random.randint(0, 1)
        src_img = np.rot90(src_img, rot)
        src_label = np.rot90(src_label, rot)
        if flip == 1:
            src_img = np.flip(src_img, 1)
            src_label = np.flip(src_label, 1)
        x_ratio = random.uniform(0.8, 1.2)
        y_ratio = random.uniform(0.8, 1.2)
```

```
        src_img = cv2.resize(src_img, (int(src_img.shape[1] * x_
ratio), int(src_img.shape[0] * y_ratio)))
                src_label = cv2.resize(src_label, (int(src_
label.shape[1] * x_ratio), int(src_label.shape[0] * y_
ratio)),
                    interpolation=cv2.INTER_NEAREST)
    # Randomly select a position
            x = random.randint(0, target_img.shape[0] - src_
img.shape[0])
            y = random.randint(0, target_img.shape[1] - src_
img.shape[1])
                target_img[x:x+src_img.shape[0], y:y+src_
img.shape[1]][src_label == choice] = src_img[src_label =
= choice]
            target_label[x:x + src_img.shape[1], y:y + src_
img.shape[1]][src_label == choice] = num_ins
    return target_img, target_label
```

3. Mosaic data augmentation

A data augmentation method used in YOLOv4 that stitches four images together to obtain a new image. The image obtained after Mosaic augmentation is shown in Fig. 6.13.

Note: Stronger data augmentation in experiments requires a larger model. Additionally, turning off this augmentation in the last 15 epochs of training can further improve the effect.[3]

---

[3] Refer to the code at https://github.com/Tianxiaomo/pytorch-YOLOv4/blob/master/dataset.py.

Fig. 6.13 Image obtained after Mosaic augmentation

References

1. Izmailov, P., Podoprikhin, D., Garipov, T., Vetrov, D., & Wilson, A. G. (2018). Averaging weights leads to wider optima and better generalization. In *Proceedings of the conference on uncertainty in artificial intelligence (UAI)*. arXiv. https://arxiv.org/abs/1803.05407
2. Solovyev, R., Wang, W., & Gabruseva, T. (2021). Weighted boxes fusion: ensembling boxes from different object detection models. *Image and Vision Computing*, 104117. https://arxiv.org/abs/1910.13302

Chapter 7
Computer Vision (Image): Practical Part

This chapter uses the Kaggle competition sartorius-cell instance segmentation (see Fig. 7.1) as an example to explain how to solve instance segmentation tasks.[1]

7.1 Competition Introduction

Neurological diseases, including neurodegenerative diseases such as Alzheimer's and brain tumors, are among the leading causes of death and disability worldwide. A feasible diagnostic method is to examine neuronal cells through optical microscopy, but due to the complex and tiny morphology of nerve cells, manual identification is very difficult and time-consuming. This competition aims to use computer vision technology to achieve automatic cell segmentation of neuronal cell microscopic images. Specifically, the task of this competition is a classic instance segmentation problem, which requires not only semantic segmentation of nerve cells in the provided microscopic images but also identification of different cells in the same image.

The training set of the competition contains 606 microscopic images of nerve cells and their corresponding ground truth (GT) labels, which are the manual segmentation results of nerve cells. The test set contains 240 images, and the test set data is not public. In addition, there are 1972 unlabeled microscopic images of nerve cells, which do not have GT for cell segmentation. All images (including labeled and unlabeled) provide information on which type of nerve cell (three types in total, namely astro, cort, shsy5y) they belong to. Examples of the three different types of nerve cells and annotation examples are shown in Fig. 7.2.

This competition uses average precision under different IoUs for evaluation, and IoU is calculated as follows.

---

[1] The competition link is https://www.kaggle.com/competitions/sartorius-cell-instance-segmentation/.

Fig. 7.1 Kaggle sartorius-cell instance segmentation competition

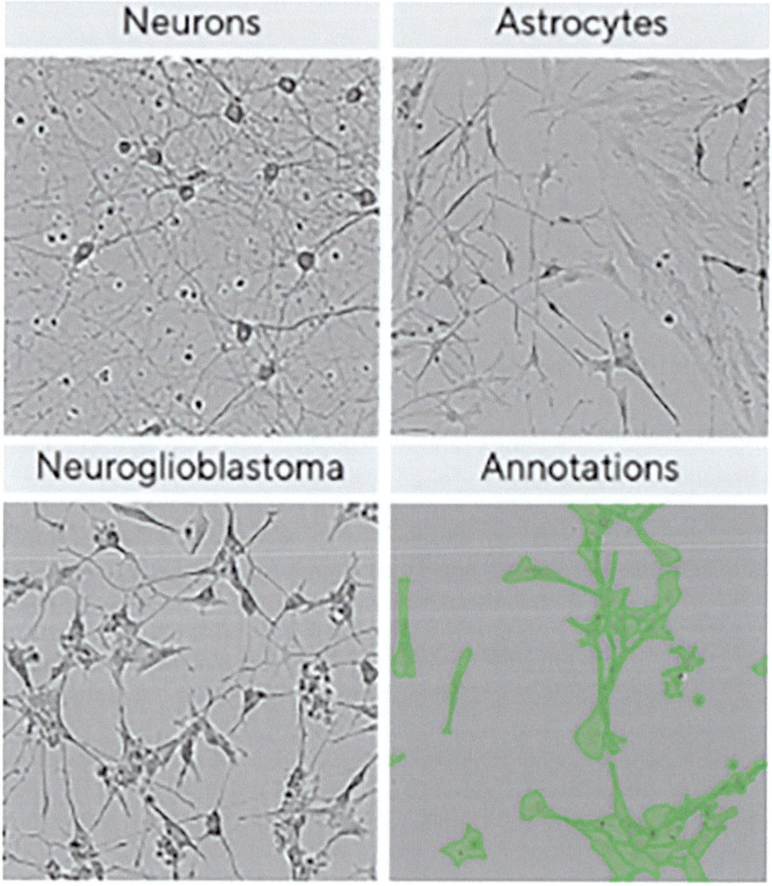

Neurons

Astrocytes

Neuroglioblastoma

Annotations

Fig. 7.2 Examples of three different types of nerve cells and annotation examples

$$\mathrm{IoU}(A, B) = \frac{A \cap B}{A \cup B}.$$

Set the IoU threshold as t, if the IoU between the predicted target and GT is greater than the threshold, the target is considered a true positive (TP), and the precision is calculated as follows.

$$\frac{\text{TP}(t)}{\text{TP}(t) + \text{FP}(t) + \text{FN}(t)}.$$

During evaluation, traverse the IoU from 0.5 to 0.95 with a step size of 0.05, calculate the precision for each IoU threshold in the above manner, and the average precision for each threshold is the score for that image. The final score is the mean of the scores for all images in the test set. The final score calculation method is as follows.

$$\frac{1}{|thresholds|} \sum_{t} \frac{\text{TP}(t)}{\text{TP}(t) + \text{FP}(t) + \text{FN}(t)}.$$

7.2 Data Exploration

This section conducts exploratory data analysis through statistical analysis, data visualization, and other means to help us better understand the data and prepare for subsequent modeling work.

7.2.1 Basic Data Situation

First, detect the statistical information of each part of the data.

The training set contains 606 images with 73,585 cell instance segmentation annotations, and the hidden test set contains approximately 240 images. In the training set, the average number of cell annotations per image is 121.42, and according to the prior information of the competition, the hidden test set has the same expected number. In addition, the train_semi_supervised directory contains 1972 images without annotations.

7.2.2 Type Distribution

There are three types of cells in the images, and each image contains only one cell type. These types include cort (neurons), shsy5y (neuroblastoma), and astro (astrocytes). Each cell type differs in characteristics and statistics, so each cell type may require its own unique processing techniques.

Figure 7.3 shows the distribution of cell types in the labeled and unlabeled training sets. It can be found that the distribution of cell types is different in the labeled and unlabeled training sets. The labeled training set has a higher number of cort, while the unlabeled training set has a higher number of astro.

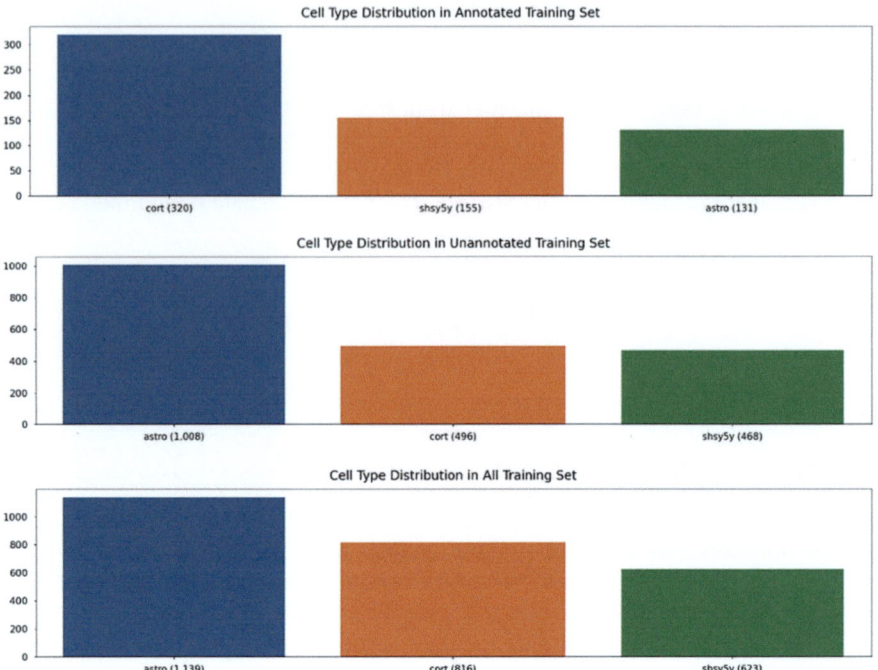

Fig. 7.3 Distribution of cell types

7.2.3　Image Distribution

The distribution of image means and standard deviations on train and train_semi_supervised is shown in Fig. 7.4. The average values and standard deviations of images in train and train_semi_supervised are slightly different, but the difference is not significant enough to determine that the images in the two datasets belong to different distributions.

The distribution of means and standard deviations for different categories of images is shown in Fig. 7.5. It can be observed that for the three different categories of cells, the differences in image means and standard deviations are quite large, with cort cells and shsy5y cells being relatively close, while the mean and standard deviation distribution of astro cells differ greatly from the former two.

7.2.4　Annotation Distribution

Figure 7.6 shows the distribution of the number and area of annotations for different categories. It can be observed that there are significant differences between different categories. The distribution of the number of cells in each cort cell image is relatively

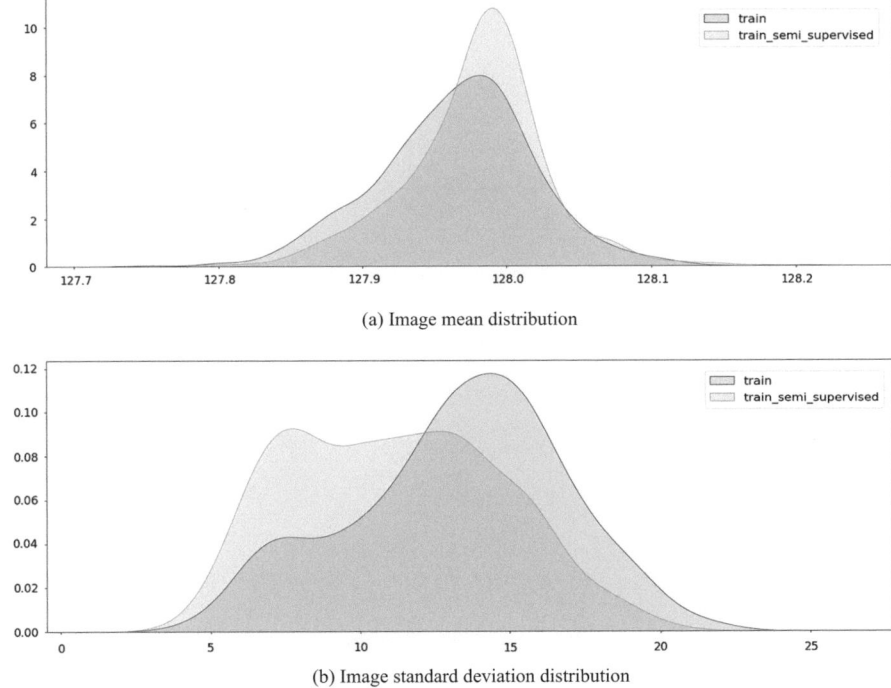

(a) Image mean distribution

(b) Image standard deviation distribution

Fig. 7.4 Distribution of image means and standard deviations on train and train_semi_supervised

concentrated, while the distribution of the number of cells in shsy5y and astro cell images is more dispersed. In terms of mask size, the area distribution of shsy5y and cort cells is very similar, indicating that the morphology of these two types of cells is very similar, while the morphology of astro cells is different from the other two types.

7.3   Interpretation of Excellent Solutions

The solution introduced in this section is based on the champion solution of the cell instance segmentation competition. The flowchart of the solution process for this competition is shown in Fig. 7.7. The overall solution consists of two stages: in the first stage, a target detection model is used to predict a bbox that can precisely enclose each target; in the second stage, a semantic segmentation model is used to segment foreground and background pixels within each bbox.

The overall solution is implemented using the MMDetection and MMSegmentation toolboxes. MMDetection and MMSegmentation are open-source object detection and semantic segmentation toolboxes based on PyTorch, supporting the training

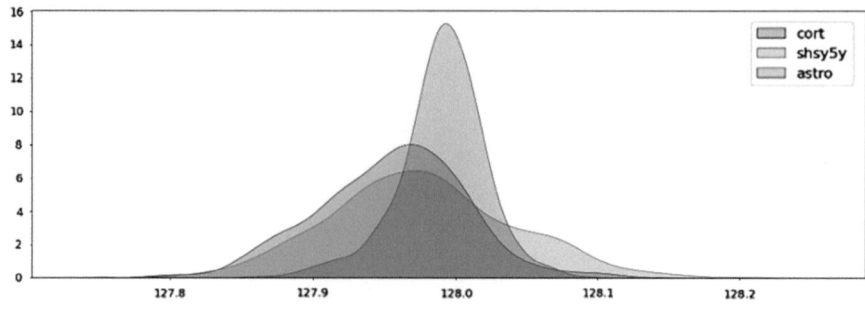

(a) Image mean distribution

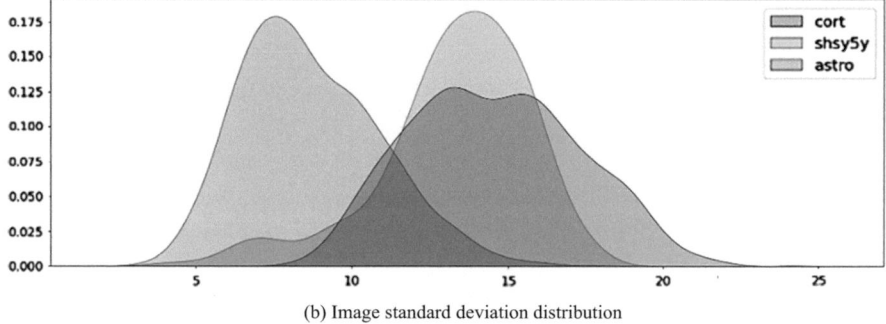

(b) Image standard deviation distribution

Fig. 7.5 Distribution of means and standard deviations for different categories of images

and deployment of various state-of-the-art models, enabling the rapid implementation of related algorithms. The versions of MMDetection[2] and MMSegmentation[3] are 2.19.0 and 0.20.2, respectively.

7.3.1 Detection Part

In this scheme, the detection part uses three YOLOX models pre-trained on the COCO dataset, trained using the GT's bbox. For the multiple bboxes predicted by the three models, the WBF method is used to integrate multiple bboxes, thereby obtaining the detection box in the first step.

1. Preprocessing

The preprocessing module mainly defines several data augmentation operations to enhance the model's generalization performance. The code for the preprocessing part is as follows.

---

[2] The link is https://github.com/open-mmlab/mmdetection.

[3] The link is https://github.com/open-mmlab/mmsegmentation.

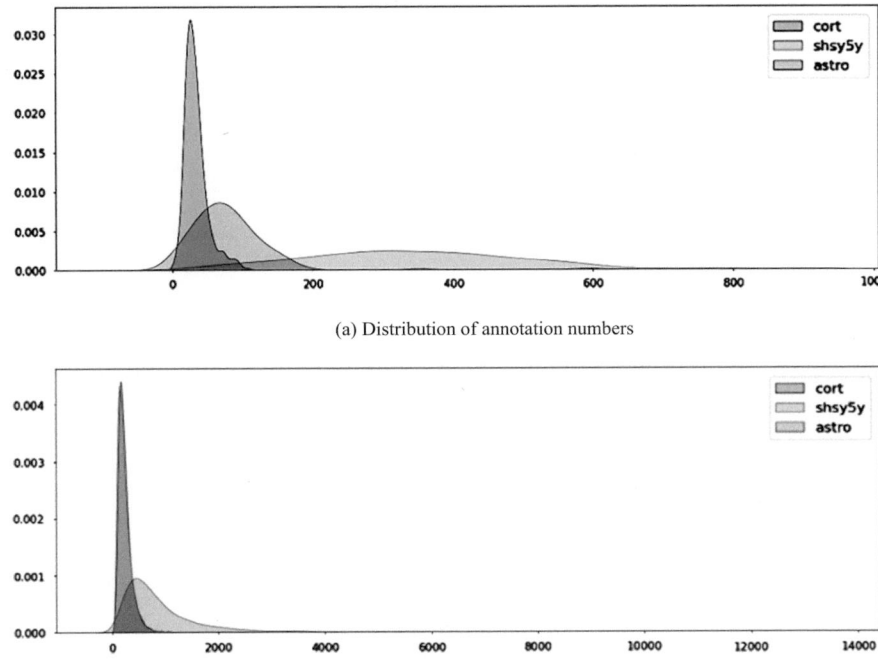

(a) Distribution of annotation numbers

(b) Distribution of annotation areas

Fig. 7.6 Distribution of the number and area of annotations for different categories

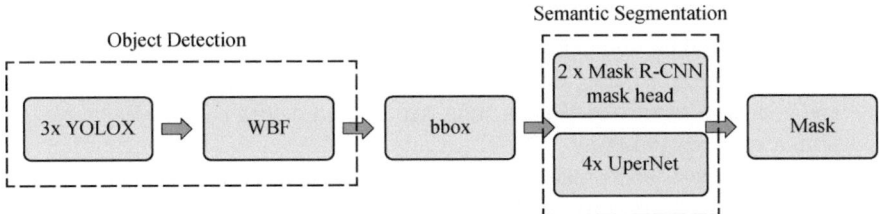

Fig. 7.7 Flowchart of the solution process for the cell instance segmentation competition

```
train_pipeline = [
    dict(type='Mosaic', img_scale=img_scale, pad_val=114.0),
    dict(
        type='RandomAffine',
        scaling_ratio_range=(0.1, 2),
        border=(-img_scale[0] // 2, -img_scale[1] // 2)
    ),
    dict(
        type='MixUp',
```

```
        img_scale=img_scale,
        ratio_range=(0.5, 1.5),
        pad_val=114.0
    ),
    dict(
        type='PhotoMetricDistortion',
        brightness_delta=32,
        contrast_range=(0.5, 1.5),
        saturation_range=(0.5, 1.5),
        hue_delta=18
    ),
    dict(type='RandomFlip', flip_ratio=0.5),
            dict(type='Resize',    img_scale=img_scale,    keep_
    ratio=True),
    dict(
        type='Pad',
        pad_to_square=True,
        pad_val=dict(img=(114.0, 114.0, 114.0))
    ),
        dict(type='FilterAnnotations', min_gt_bbox_wh=(1, 1),
    keep_empty=False),
    dict(type='DefaultFormatBundle'),
        dict(type='Collect', keys=['img', 'gt_bboxes', 'gt_
    labels'])
    ]
```

Each "dict" in the above code is a preprocessing step, where the "type" field indicates the type of this step. The code analysis is as follows.

(1) The Mosaic data augmentation method is used to combine multiple images through random scaling, cropping, and arrangement, enlarging the image to a specified size (controlled by the "img_scale" parameter), and padding the edges with a value of 114.0.

(2) Random affine transformations are applied to the image (i.e., randomly applying rotation, translation, scaling, shearing, flipping to the image), with a scaling ratio range of (0.1, 2), and setting the edge padding value.

(3) The mixup operation is used to mix two images together, resizing them to "img_scale," with a mixing ratio range of (0.5, 1.5), and a padding value of 114.0.

(4) Random photometric distortion (PhotoMetricDistortion) is applied to the image, i.e., random distortion of brightness, contrast, saturation, and hue, with a brightness offset of 32, contrast range of (0.5, 1.5), saturation range of (0.5, 1.5), and hue offset of 18.

(5) The Resize operation is used to scale the image to the specified size while maintaining the aspect ratio.

(6) The image is expanded into a square, and the surrounding blank areas are filled with a specified color.

(7) Bounding boxes smaller than a certain threshold are filtered.

(8) Default formatting (DefaultFormatBundle) is applied to the image and annotation information.

(9) Image, bounding box, and label information are extracted from the raw data as input to the model.

2. Model

The author used the YOLOX network structure (see Fig. 7.8) as the detection model and trained three identical network structures with different random seeds for ensemble learning. YOLOX is an improved version of the YOLO series open-sourced by Megvii in 2021, achieving good results by introducing methods such as Decoupled Head and SimOTA on the basis of YOLO v3.

The configuration information of the YOLOX model is as follows.

```
model = dict(
    type='YOLOX',
    input_size=img_scale,
    random_size_range=(32, 64),
    random_size_interval=1,
        backbone=dict(type='YOLOPAFPNOfficial', depth=1.33,
width=1.25),
    neck=None,
    bbox_head=dict(
        type='YOLOXHeadOfficial',
        num_classes=3,
        width=1.25,
        in_channels=[256, 512, 1024]
    ),
        train_cfg=dict(assigner=dict(type='SimOTAAssigner',
center_radius=2.5)),
    test_cfg=dict(score_thr=0.01, nms=dict(type='nms', iou_
threshold=0.65))
)
```

The configuration information is parsed as follows.

The input image size for the model is img_scale. During model training, the input image size is randomly cropped, and the cropped image is used as the model's input. The random_size_range indicates the range of random cropping sizes, and random cropping is performed every random_size_interval iterations. The model's backbone is set to YOLOPAFPNOfficial, and the bbox head type is YOLOXHeadOfficial. The relevant code implementations are in cell_modules. Since this case involves 3 similar cells, the number of classes, num_classes, is set to 3.

During training, the SimOTAAssigner allocator is used to assign GT boxes to the model's predicted boxes. This allocator uses a center radius to determine which model prediction boxes each GT box can be assigned to. Only prediction boxes with a center distance to the GT box center of less than 2.5 can be assigned. This method helps the model learn better how to predict object locations.

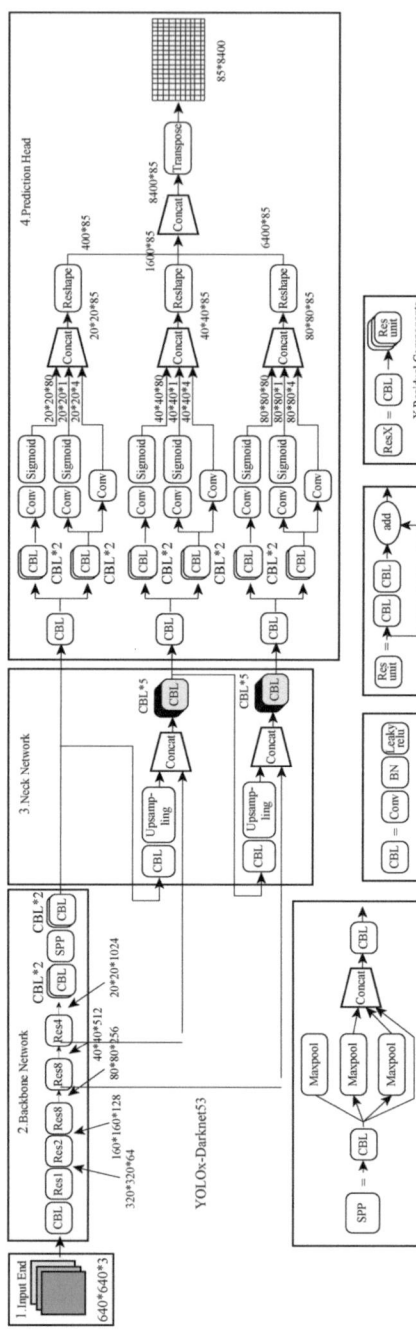

Fig. 7.8 YOLOX network structure diagram

During testing, the NMS algorithm is used for post-processing, with an IoU threshold of 0.65.

3. Training details

The training details include setting the optimizer type, learning rate, momentum, and other parameters, and controlling the optimizer's behavior through some custom hook functions. The specific configuration parameters are as follows.

```
optimizer = dict(
    type='SGD',
    lr=0.005 / 64,
    momentum=0.9,
    weight_decay=0.0005,
    nesterov=True,
        paramwise_cfg=dict(norm_decay_mult=0.0, bias_decay_
mult=0.0)
)
lr_config = dict(
    policy='YOLOX',
    warmup='exp',
    by_epoch=False,
    warmup_by_epoch=True,
    warmup_ratio=1,
    warmup_iters=3,
    num_last_epochs=5,
    min_lr_ratio=0.01
)
custom_hooks = [
        dict(type='YOLOXModeSwitchHook', num_last_epochs=15,
priority=48),
    dict(
        type='ExpMomentumEMAHook',
        resume_from=resume_from,
        momentum=0.0002,
        total_iter=500,
        priority=49
    )
]
runner = dict(type='EpochBasedRunner', max_epochs=10)
```

The configuration parameters are parsed as follows.

The optimizer defines the optimizer configuration, with the optimizer type as SGD, an initial learning rate set to 0.005/64, momentum of 0.9, weight decay rate of 0.0005, and using Nesterov momentum.

The lr_config defines the learning rate adjustment strategy, using the same strategy as in the YOLOX paper, which is the warmup + cosine annealing strategy. Additionally, a fixed minimum learning rate is used for the last 5 epochs.

Finally, two custom hook functions are defined. The YOLOXModeSwitchHook is used to disable Mosaic and mixup data augmentation in the last 15 epochs and add additional L1 loss. The EMA strategy is used to perform exponential averaging on the final model. The total training epochs are 30.

4. Ensemble learning

Finally, the WBF method is used to fuse the detection boxes predicted by multiple models to obtain the final result. The ensemble learning code is as follows.

```
from ensemble_boxes import weighted_boxes_fusion
boxes, scores, labels = weighted_boxes_fusion(boxes_list,
scores_list, labels_list, iou_thr=0.5, skip_box_thr=0.0001)
```

7.3.2 Segmentation Part

The segmentation part uses two Mask R-CNN Mask heads and four UperNet models for foreground and background segmentation. Among them, Mask R-CNN uses CB-DBS as the backbone, and UperNet uses Swin Transformer and ResNet-101 as backbone networks. Training is performed using bounding boxes and masks from GT, and inference uses boxes obtained from the output of the previous detection part.

Due to space limitations, we only show the implementation of one UperNet model here.

1. Preprocessing

The preprocessing workflow defines how the data used for training is preprocessed, including operations such as random shifting, alignment, flipping, and rotating of images, as well as image normalization. These preprocessing operations can enhance the model's generalization ability, improving its performance on the test set. The configuration information for the preprocessing section is as follows.

```
train_pipeline = [
   dict(type='BoxJitter', prob=0.5),
   dict(type='ROIAlign', output_size=crop_size),
   dict(type='FlipRotate'),
   dict(type='Normalize', **img_norm_cfg),
   dict(type='DefaultFormatBundle'),
   dict(type='Collect', keys=['img', 'gt_semantic_seg'])
]
img_norm_cfg = dict(
      mean=[123.675, 116.28, 103.53], std=[58.395, 57.12,
57.375], to_rgb=True
   )
```

The above code defines the data preprocessing workflow in MMDetection. The code is explained as follows.

(1) BoxJitter: Randomly shifts the bounding boxes in the image with a 50% probability to simulate potential inaccuracies in detection boxes during prediction.
(2) ROIAlign: Since the general target scale is small, the ROIAlign method is used to align the image and GT mask, with the output size being "crop_size." The GT mask is rounded to the nearest integer after bilinear interpolation.
(3) FlipRotate: Randomly flips or rotates the image.
(4) Normalize: Normalizes the image using mean and standard deviation, and converts it to RGB format. Specific parameters are defined in "img_norm_cfg."
(5) DefaultFormatBundle: Formats the image and annotation information in the default manner.
(6) Collect: Collects the required data, organizing images and labels into batches.

2. UperNet Model

UperNet (see Fig. 7.9 [1]) is a unified perceptual parsing network published by Megvii at ECCV18. It manually designed a Broden + dataset and defined a recognition task called unified perceptual parsing (UPP), attempting to parse multi-level visual concepts of an image from scene, object, part, material to texture in one go. A multi-task network and a training strategy for handling mixed annotations were developed and tested, further utilizing the trained network to discover visual knowledge in the scene.

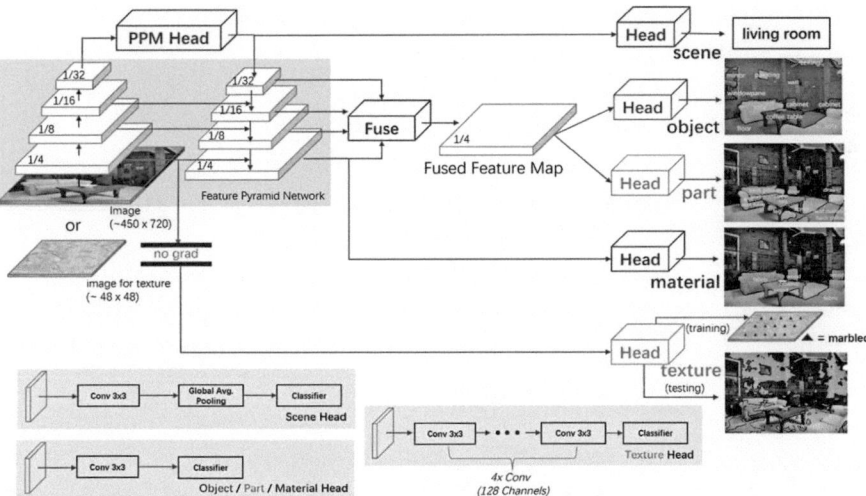

Fig. 7.9 UperNet schematic

The model configuration file is as follows.

```
model = dict(
    type='CustomEncoderDecoder',
    backbone=dict(
        type='SwinTransformer',
        pretrain_img_size=224,
        embed_dims=96,
        patch_size=4,
        window_size=7,
        mlp_ratio=4,
        depths=[2, 2, 6, 2],
        num_heads=[3, 6, 12, 24],
        strides=(4, 2, 2, 2),
        out_indices=(0, 1, 2, 3),
        qkv_bias=True,
        qk_scale=None,
        patch_norm=True,
        drop_rate=0.0,
        attn_drop_rate=0.0,
        drop_path_rate=0.3,
        use_abs_pos_embed=False,
        act_cfg=dict(type='GELU'),
        norm_cfg=backbone_norm_cfg
    ),
    decode_head=dict(
        type='UPerHead',
        in_channels=[96, 192, 384, 768],
        in_index=[0, 1, 2, 3],
        pool_scales=(1, 2, 3, 6),
        channels=512,
        dropout_ratio=0.1,
        num_classes=2,
        norm_cfg=norm_cfg,
        align_corners=False,
        loss_decode=dict(
            type='CrossEntropyLoss', use_sigmoid=False, loss_
weight=1.0
        )
    ),
    auxiliary_head=dict(
        type='FCNHead',
        in_channels=384,
        in_index=2,
        channels=256,
        num_convs=1,
        concat_input=False,
        dropout_ratio=0.1,
        num_classes=2,
        norm_cfg=norm_cfg,
        align_corners=False,
```

```
        loss_decode=dict(
             type='CrossEntropyLoss', use_sigmoid=False, loss_
    weight=0.4
        )
    ),
    train_cfg=dict(),
    test_cfg=dict(mode='whole')
)
```

The model consists of a custom encoder–decoder (CustomEncoderDecoder). The model includes the SwinTransformer backbone and two heads, where one head (decode_head) is used for predicting pixel segmentation, and the other head (auxiliary_head) is used to assist in training to predict cell types.

3. Training details

The optimizer and learning rate adjustment strategy used during model training are as follows.

```
optimizer = dict(
    type='AdamW',
    lr=6e-5 / 16,
    betas=(0.9, 0.999),
    weight_decay=0.01,
    paramwise_cfg=dict(
        custom_keys=dict(
            absolute_pos_embed=dict(decay_mult=0.0),
            relative_position_bias_table=dict(decay_mult=0.0),
            norm=dict(decay_mult=0.0)
        )
    )
)
lr_config = dict(
    policy='CosineAnnealing',
    by_epoch=False,
    warmup='linear',
    warmup_iters=100,
    warmup_ratio=1.0 / 10,
    min_lr_ratio=1e-5
)
runner = dict(type='EpochBasedRunner', max_epochs=10)
```

Specifically, the above code uses AdamW as the optimizer, with an initial learning rate of $6e^{-5}/16$ and a weight decay set to 0.01.

The learning rate adjustment strategy uses linear warmup and CosineAnnealing, with a total of 10 training epochs.

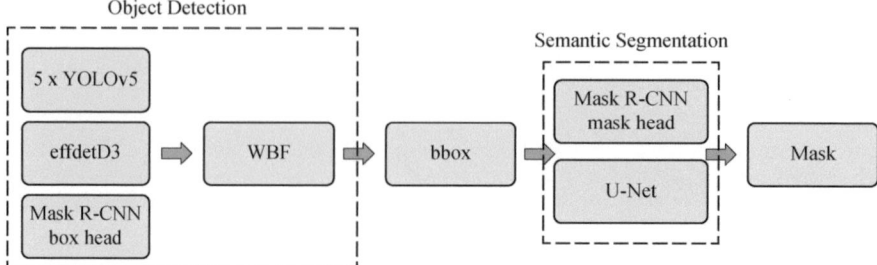

Fig. 7.10 Runner-up solution flowchart

4. Ensemble learning

The ensemble learning module averages the pixel segmentation probabilities output by multiple models to obtain the final predicted mask.[4]

7.4   More Solutions

1. Runner-up solution

Figure 7.10 shows the flowchart of the runner-up solution. The overall process of the runner-up solution is similar to the champion solution. The bbox part uses 5 YOLOv5, effdetD3, and Mask R-CNN's box head, while the Mask part uses U-Net and Mask R-CNN's mask head. It is worth mentioning that, as shown in Fig. 7.11, in the runner-up solution, the mask part is not based on the feature map but directly crops small image patches from the original image according to the bbox and inputs the patches directly into the segmentation network for foreground and background segmentation. As shown in Fig. 7.12, to avoid confusion in the segmentation of adjacent cells, the segmentation task for each bbox is threefold, namely target cell, other cells, and background.

In Fig. 7.12, from left to right are the cropped original image, target cell annotation, target cell prediction, other cell annotation, and other cell prediction.

2. Third-place solution

The flowchart of the third-place solution is shown in Fig. 7.13.

This solution is an ensemble of three schemes, namely category-agnostic Mask R-CNN, category-specific Mask R-CNN, and Cellpose, the integration strategy is to first cluster the masks belonging to the same target based on IoU, and then for each target, average all the masks belonging to that category. The category-specific

---

[4] The complete code implementation can be found at https://github.com/enjoysport2022/DataMi ningCompetitionInAction/.

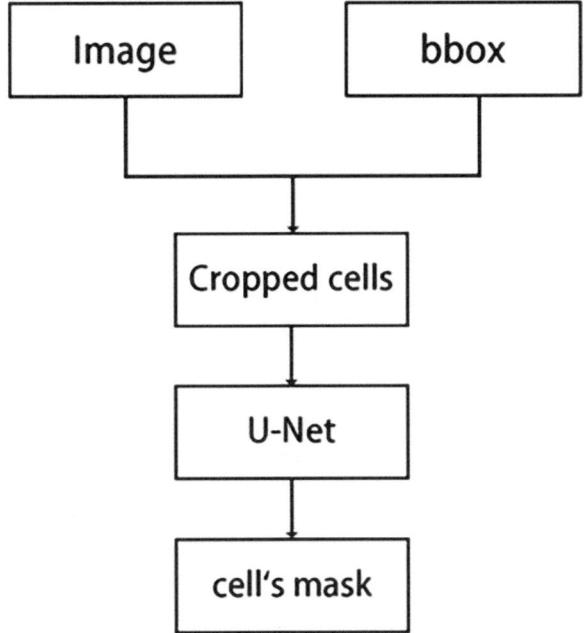

Fig. 7.11 Runner-up solution segmentation process

Fig. 7.12 Segmentation task illustration

Fig. 7.13 Third-place solution flowchart

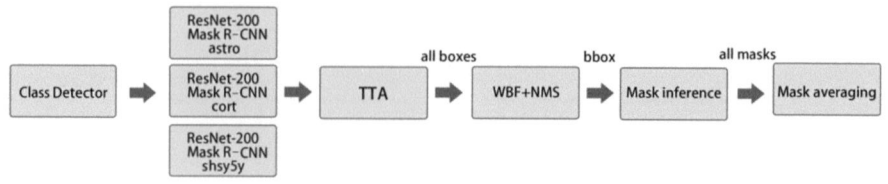

Fig. 7.14   Model process for distinguishing cell types

Mask R-CNN is shown in Fig. 7.14. First, a classification network, called class detector, is trained to distinguish the cell types in the image (three different types of neural cells: astro, cort, shsy5y). Then, different Mask R-CNNs are used for instance segmentation according to the type. In this scheme, TTA is used to improve accuracy, which means applying different transformations to the input image during testing to generate multiple copies. Each copy is independently predicted, and the outputs from multiple copies are integrated in both the detection and segmentation parts. WBF + NMS is used to integrate the bbox of multiple input copies, and then the average method is used to integrate the masks.

Reference

1. Xiao, T., Liu, Y., Zhou, B., Jiang, Y., & Sun, J. (2018). *Unified perceptual parsing for scene understanding* [Preprint]. arXiv. https://arxiv.org/pdf/1807.10221.pdf

Chapter 8
Computer Vision (Video): Theoretical Part

8.1 Differences Between Video Data and Image Data

Video data differs from static image data in that it is a series of dynamic image sequences ordered by time. In video data, there is a close contextual relationship between frames, so it is necessary to consider motion correlation and temporal correlation. This means that when modeling, it is necessary to focus on both the appearance information and motion information of the data.

Compared to image data, which only contains RGB information, video data has the following modalities.

1. RGB Information

As shown in Fig. 8.1, the RGB format of video data is generally (H, W, C, T), which represents the height of the image, the width of the image, the number of channels, and the number of frames. This is the most basic data representation of video. There is a causal and sequential relationship between the frames of a video, for example, the actions of "opening a door" and "closing a door" need to be distinguished through temporal information.

Currently, mainstream deep learning-based video understanding models generally rely solely on RGB information end-to-end modeling.

2. Optical Flow Calculation

As shown in Fig. 8.2 [1], the optical flow algorithm performs additional processing on sequential RGB images to obtain motion estimation. It estimates the direction and speed of object motion by comparing the positions of pixels between consecutive frames. This algorithm can estimate the motion of objects or backgrounds in an image and is often used for tracking and motion analysis.

Common optical flow algorithms include Lucas-Kanade, Horn-Schunck, and Farneback, each having different advantages and disadvantages in various scenarios.

© Tsinghua University Press 2026
K. Xu, *Data Mining Competition Practices*,
https://doi.org/10.1007/978-981-95-3446-3_8

Time

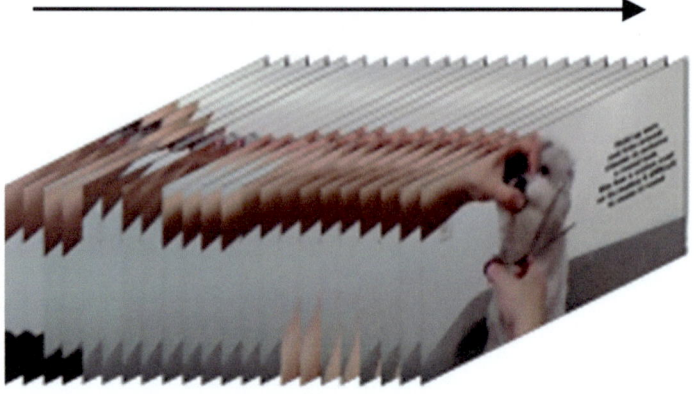

Fig. 8.1 RGB information of video data

The advantage is that dual-stream models combining optical flow information and RGB information can perform well on smaller datasets, with relatively low inference overhead; however the disadvantage is that optical flow algorithms introduce additional overhead and cannot provide end-to-end training and inference scenarios.

The following is an example code for the classic optical flow algorithm Lucas-Kanade implemented in OpenCV.

```python
import numpy as np
import cv2
import sys
cap = cv2.VideoCapture("video.mp4")
feature_params = dict(maxCorners = 100, qualityLevel = 0.3,
minDistance = 7, blockSize = 7)
lk_params   =   dict(winSize=(15,   15),   maxLevel   =   2,
criteria=(cv2.TERM_CRITERIA_EPS | cv2.TERM_CRITERIA_COUNT,
10, 0.03))
color = np.random.randint(0, 255, (100, 3))
ret, old_frame = cap.read() # Read the first frame
old_gray = cv2.cvtColor(old_frame, cv2.COLOR_BGR2GRAY)
p0   =   cv2.goodFeaturesToTrack(old_gray,   mask   =   None,
**feature_params) # Select feature points and return the
list of feature points
mask = np.zeros_like(old_frame)
while(1):
    ret, frame = cap.read() # Read the next frame
    if frame is None:
        break
    frame_gray = cv2.cvtColor(frame, cv2.COLOR_BGR2GRAY)
```

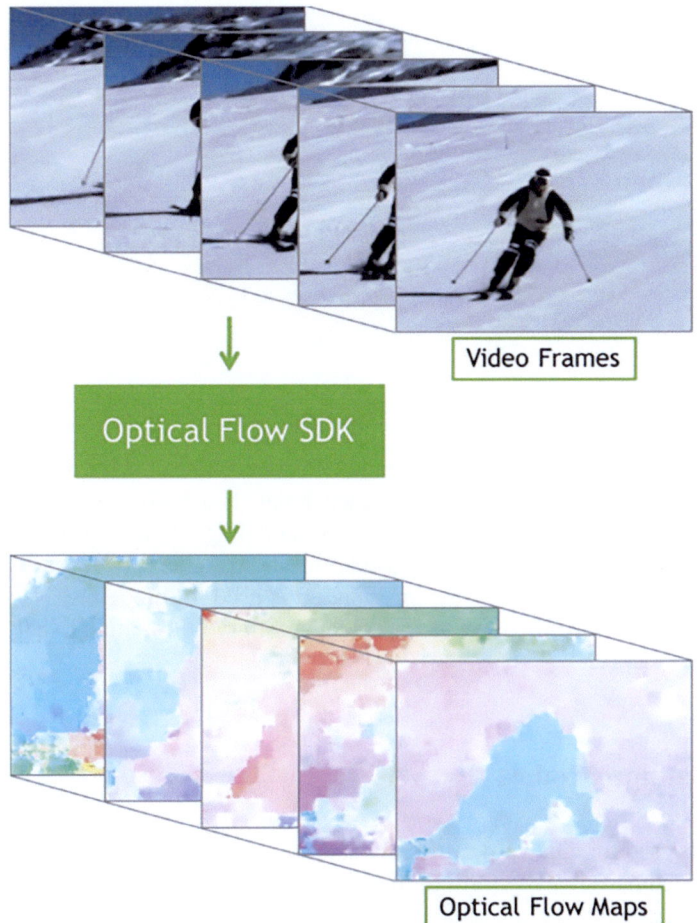

Fig. 8.2 Optical flow algorithm

```
    p1, st, err = cv2.calcOpticalFlowPyrLK(old_gray, frame_
gray, p0, None, **lk_params) # Calculate the positions of the
corresponding feature points in the new image
    good_new = p1[st == 1]
    good_old = p0[st == 1]
    for i, (new, old) in enumerate(zip(good_new, good_old)):
        a, b = new.ravel() # Flatten the array into a one-
dimensional array
    c, d = old.ravel()
    mask = cv2.line(mask, (a, b), (c, d), color[i].tolist(),
2)
```

```
        frame = cv2.circle(frame, (a, b), 5, color[i].tolist(),
 -1) # Draw the optical flow visualization image
    img = cv2.add(frame, mask)
    cv2.imshow('frame', img)
    k = cv2.waitKey(30) & 0xff
    if k == 27:
        break
    old_gray = frame_gray.copy()
    p0 = good_new.reshape(-1, 1, 2)
cv2.destroyAllWindows()
cap.release()
```

3. Audio Information

In video data, audio information is often included, and this audio information usually needs to be converted first into two-dimensional data before being input into a convolutional neural network. This enables modeling of the audio information. However,except in some multimodal scenarios, audio signals are generally not used in competitions, where only visual information is used for modeling. Nevertheless, audio information can still serve as important auxiliary information to help improve model performance.

8.2   Common Models

The models used for video tasks differ from those for static images and can generally be categorized based on the form of input.

1. CNN+RNN

This type of network is represented by long-term recurrent convolutional networks (LRCN). The structure diagram of LRCN is shown in Fig. 8.3 [1]. First, a single frame image is convolved to extract spatial features, then a recurrent neural network (such as LSTM) is used to seqentially model the extracted single frame image features, and finally, the final prediction result of the entire video segment is obtained.

The advantages and disadvantages of the LRCN structure are as follows.

Advantages: It can leverage the weights pre-trained on large-scale datasets (such as ImageNet) by convolutional neural networks.

Disadvantages: The RNN in the second stage can only model the high-level semantic features extracted by the CNN and lacks the ability to capture low-level motion features.

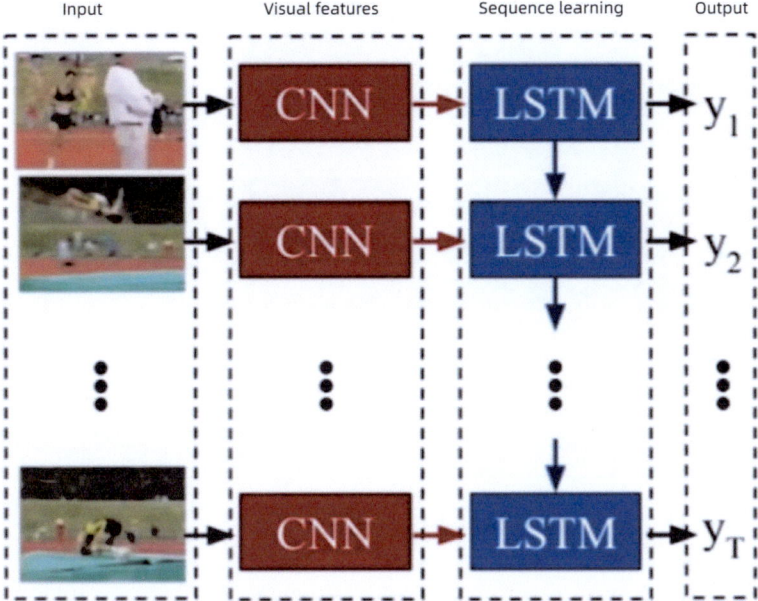

Fig. 8.3 Structure diagram of LRCN

The LRCN series of networks may perform better on datasets with significant differences in single-frame image information. The model has the fewest parameters among the three basic models.[1]

2. Two-Stream Network

The two-stream network explicitly segments the video information stream, modeling the frame information stream and motion information stream separately. Most two-stream networks use optical flow to represent motion information and RGB data of single-frame images to represent frame information. For both types of information, pre-trained convolutional networks are used to extract features and make independent predictions, then the independent prediction results of different information streams are fused to obtain the final classification prediction of the video segment. Figure 8.4 [2] shows the structure of the classic two-stream network, the Temporal Segment Network (TSN), where the Spatial and Temporal branches independently process frame information and motion information.

---

[1] LRCN implementation code can be found at https://github.com/garythung/torch-lrcn.

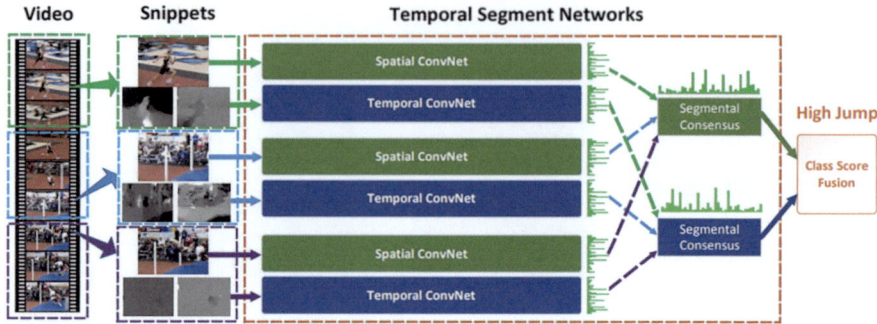

Fig. 8.4 Schematic diagram of TSN structure

| Conv1a | | Conv2a | | Conv3a | Conv3b | | Conv4a | Conv4b | | Conv5a | Conv5b | | fc6 | fc7 | |
| 64 | | 128 | | 256 | 256 | | 512 | 512 | | 512 | 512 | | 4096 | 4096 | |

Fig. 8.5 Network structure diagram of C3D

The advantages and disadvantages of the TSN[2] structure are as follows.

Advantages: It can naturally leverage most pre-trained CNN networks, such as VGG, ResNet, etc., achieving good results. Currently, many studies are based on two-stream networks. Optical flow information models motion information a priori and can achieve good results even in scenarios with relatively small sample sizes.

Disadvantages: Most works require pre-training on optical flow information, which does not support an end-to-end pre-training, and optical flow calculation consumes significant computational resources.

3.   3D Convolutional Network

The 3D convolutional network, with classic work such as C3D, extends the two-dimensional convolution kernel into the time dimension, expanding the convolution kernel from a plane to a cube, naturally forming a three-dimensional convolution kernel.[3] By stacking 3D convolution blocks, it learns spatiotemporal features in video data in an end-to-end manner. Figure 8.5 [3] shows the network structure diagram of C3D, which captures the ability to acquire spatiotemporal features through stacked 3D convolution blocks. 3D convolutional networks have broad applications in high-dimensional medical imaging and video analysis, with many attempts to decompose 3D convolution into 2D+1D convolution operations, all with good results. This model has the largest number of parameters among the three basic models.

---

[2] The TSN implementation code is available at https://github.com/yjxiong/temporal-segment-net works.

[3] The implementation code for the classic 3D convolutional network SlowFast can be found at https://github.com/facebookresearch/SlowFast.

The advantages and disadvantages of 3D convolutional networks are as follows.

Advantages: It is a natural extension of 2D convolution, learning motion information and frame information simultaneously, theoretically achieving true spatiotemporal semantic extraction.

Disadvantages: The parameter count is cubic, leading to a large number of parameters and a tendency to overfit. It cannot directly utilize pre-trained models on image datasets to initialize model parameters. This issue has been alleviated after applications on large-scale video datasets like Kinetics.

4. 3D Transformer

The 3D Transformer network extends self-attention by one dimension in the temporal direction, forming three-dimensional self-attention to achieve end-to-end learning of the spatiotemporal semantic features of videos. Due to the large number of parameters introduced by spatiotemporal global self-attention, different models adopt different spatiotemporal modeling strategies.

The advantages and disadvantages of the 3D Transformer are as follows.

Advantages: Similar to 3D convolutional networks, it can learn motion information and image information in a one-stop manner, while having the characteristics of self-attention, overcoming the locality of convolution, and modeling spatiotemporal global interactions at each layer.

Disadvantages: The number of parameters is larger compared to 3D convolutional networks, which can easily lead to overfitting. It requires pre-training on a large amount of data and cannot be directly utilized on image datasets.

The 3D Transformer is currently the most popular model architecture and can be used as a base model for exploration. Representative methods include VTN (Video Transformer Network), ViViT (A Video Vision Transformer), TimeSformer,[4] Video-Swin-Transformer,[5] MViT (Multiscale Vision Transformer),[6] etc. Figure 8.6 [4] shows the spatiotemporal modeling methods in TimeSformer. The first column is the standard ViT attention form without considering temporal information; the second column is the spatiotemporal global attention form that calculates all temporal patches simultaneously; the third column first calculates the attention of patches at the same position between frames, then calculates the attention of all patches within a frame, which is the most effective form in TimeSformer experiments.

---

[4] The implementation code for TimeSformer can be found at https://github.com/facebookresearch/TimeSformer.

[5] The implementation code for Video-Swin-Transformer can be found at https://github.com/SwinTransformer/Video-Swin-Transformer.

[6] The implementation code for MViT can be found at https://github.com/facebookresearch/mvit.

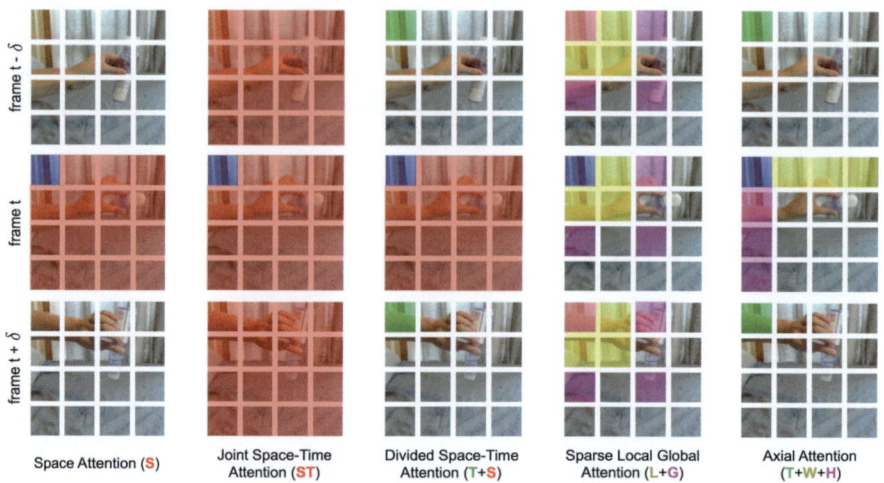

Fig. 8.6 Three spatiotemporal modeling methods proposed in TimeSformer

8.3   Pre-training Datasets

Table 8.1 lists some public video datasets and basic information. You can choose a suitable dataset for pre-training based on your computing power and task type. Among these, Kinetics is the most commonly used pre-training dataset.

8.4   Task Introduction

Video understanding involves multiple aspects of tasks, and it has now developed into a very broad academic research and industrial application direction. Due to space limitations, here we provide a brief introduction to several basic tasks in video understanding: action recognition (video classification), temporal action localization, spatiotemporal action detection, and video object detection.

1.   Action Recognition (Video Classification)

The goal of action recognition is to identify actions appearing in a video, usually human actions, which can be seen as a natural extension from the field of image classification to the video domain. As shown in Fig. 8.7, each video clip corresponds to a unique label, and our goal is to recognize the person riding a bicycle in the video.

2.   Temporal Action Localization

Action recognition can be viewed as a classification problem, where the videos to be recognized are typically edited, meaning each video contains only one clear action, the video duration is relatively short, and there is a uniquely determined action

Table 8.1 Public video datasets and basic information

Dataset	Basic task	Number of categories	Total scale	Average duration (s)	Total duration (h)
HMDB51	Action recognition	51	6714	3–10	–
UCF101	Action recognition	101	13320	7.21	26.67
ActivityNet1.3	Action recognition	200	20000	180	700
Charades	Action recognition	157	9848	–	–
Kinetics400	Action recognition	400	236532	10	657
Kinetics-Sounds	Action recognition	31	18716	10	51
EPIC-KITCHENS-100	Action recognition	v.97, n.300	89977	3.1	100
THUMOS'14	Temporal localization	20	413	68.86	7.56
AVE	Audio-visual localization	28	4143	10	11
LLP	Audio-visual localization	25	11849	10	33
AVSBench	Audio-visual segmentation	23	4932	5	6.85
VGGsound	Action recognition	309	185229	10	514
MUSIC-AVQA	Audio-visual question answering	22	9288	60	150
Breakfast	Action segmentation	1712	1989	139.37	77
50Salads	Action segmentation	17	50	384	5.33
GTEA	Action segmentation	7	28	74.34	0.58
EGTEA Gaze++	Temporal localization, etc.	106	86	1214	29
Ego4D	Temporal localization, etc.	–	–	–	3670

riding a bike

Fig. 8.7 Action recognition

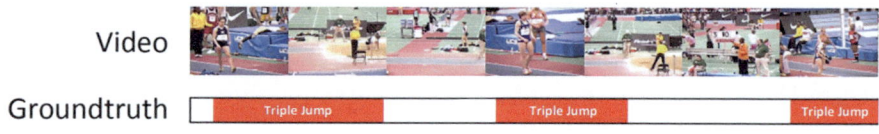

Fig. 8.8 Temporal action localization

category. In the field of temporal action localization, however, videos are usually unedited, the duration is longer, actions typically occur only in a small segment of the video, and the video may contain multiple actions or none at all, in which case the video is referred to as a background class. As shown in Fig. 8.8, temporal action localization not only predicts what actions are contained in the video but also predicts the start and end times of the actions. Compared with action recognition, temporal action localization is closer to real-world scenarios.

If each frame of the image is regarded as a token, temporal action localization is similar to the named entity recognition task in the NLP field, requiring the identification of entities appearing in frames and locating the start and end frames. It can be directly modeled as a frame-level classification task, or techniques from named entity recognition like GlobalPointer can be referenced.

3. Spatiotemporal Action Detection

Spatiotemporal action detection requires identifying the action categories in a video and determining their time and location, necessitating the simultaneous detection and localization of specific actions (such as running, jumping, etc.) in the video and assigning a timestamp (i.e., the start and end time of the action). This task combines elements of action recognition, object detection, and temporal processing, requiring spatial and temporal analysis of the video. This task has important applications in the fields of computer vision and artificial intelligence, such as intelligent surveillance, sports analysis, and video search.

This task is similar to video object detection, but since actions need to be judged through temporal information, static object detection methods are generally not used.

Instead, 3D backbones suitable for video modeling are used to integrate detection models (such as RCNN).

4. Video Object Detection

Video object detection is the application of object detection tasks on video data, with task definitions identical to traditional object detection. In some scenarios, such as image blur, occlusion, or unusual target poses, using static images alone for object detection often yields poor results. By using features from other frames, prediction effects can be enhanced, and contextual information can help recognize action information.

Depending on the usage scenario and task, video object detection problems can be approached using the following methods.

(1) Post-processing Based

Perform object detection on each frame's static image separately, then dynamically correct based on inter-frame information, such as sequence non-maximum suppression (Seq-NMS), sequence box matching (Seq-Bbox Matching), REPPx.[7] This approach has high detection response, can be applied to real-time detection, and is flexible, allowing integration with any object detection algorithm.

(2) Tracking Based

DeepSORT uses Kalman filtering from control theory to predict the motion of targets, then uses the Hungarian algorithm to match the predicted positions with the targets detected in a new frame. The process follows: The detector obtains bbox $\rightarrow$ generates a detection $\rightarrow$ The Kalman filter predicts the track $\rightarrow$ The Hungarian algorithm matches the predicted tracks with the detections in the current frame $\rightarrow$ Updates track the prediction.

(3) 3D CNN (Transformer)

Use 3D networks for end-to-end video object detection, such as YOLOV, TransVOD.[8]

References

1. Donahue, J., Hendricks, L. A., Rohrbach, M., Venugopalan, S., Guadarrama, S., Saenko, K., & Darrell, T. (2014). *Long-term recurrent convolutional networks for visual recognition and description* [Preprint]. arXiv. https://doi.org/10.48550/arXiv.1411.4389
2. Wang, L., Xiong, Y., Wang, Z., Qiao, Y., Lin, D., Tang, X., & Van Gool, L. (2016). Temporal segment networks: towards good practices for deep action recognition. In B. Leibe, J. Matas, N. Sebe, & M. Welling (Eds.), *Computer vision—ECCV 2016* (pp. 20–36). Springer. https://doi.org/10.1007/978-3-319-46484-8_2

[7] The implementation code of REPP can be found at https://github.com/AlbertoSabater/Robust-and-efficient-post-processing-for-video-object-detection.

[8] The implementation code of TransVOD can be found at https://github.com/SJTU-LuHe/TransVOD.

3. Tran, D., Bourdev, L., Fergus, R., Torresani, L., & Paluri, M. (2015). Learning spatiotemporal features with 3D convolutional networks. In *Proceedings of the IEEE international conference on computer vision (ICCV)* (pp. 4489–4497). https://arxiv.org/abs/1412.0767
4. Bertasius, G., Wang, H., & Torresani, L. (April 16, 2021). Is Space-Time Attention All You Need for Video Understanding? [cs.CV]. arXiv. https://arxiv.org/pdf/2102.05095v3

Chapter 9
Computer Vision (Video): Practical Part

This chapter uses the competition "PRE-TRAINING FOR VIDEO UNDER-STANDING CHALLENGE" held by ACM Multimedia 2022 (see Fig. 9.1) as an example to explain practical cases of video understanding competitions.

9.1 Background of the Competition

In recent years, with the rise of the short video field, there are billions of multimedia videos on the internet, which often have weak labels such as video titles and categories, characterized by high labeling noise and large category spans. Although the latest advances in computer vision have achieved considerable success in areas such as video classification, video captioning, and video object detection, how to effectively utilize the vast amount of unlabeled or weakly labeled videos widely available on the internet remains a topic worth researching. This "PRE-TRAINING FOR VIDEO UNDERSTANDING CHALLENGE" aims to promote research on video pre-training technology and encourage research teams to design new pre-training techniques to enhance a series of downstream tasks.

9.2 Data Introduction and Evaluation Metrics

The competition provides a dataset of 3 million videos, YOVO-3M, scraped from YouTube (see Fig. 9.2, image source: competition homepage) for model pre-training. Each video includes a title from YouTube and a query representing the video category (such as bowling, archery, tiger cat, etc.).

The competition also provides a downstream task dataset, YOVO-downstream, containing about 100,000 videos. This dataset includes a training set of 70,173 videos,

© Tsinghua University Press 2026
K. Xu, *Data Mining Competition Practices*,
https://doi.org/10.1007/978-981-95-3446-3_9

Fig. 9.1   Pre-training for video understanding challenge

Query: brushing teeth   **Query:** bowling   **Query:** archery   **Query:** tiger cat

Sentence: Disney Jr Puppy Dog Pals   **Sentence:** Dude Perfect   **Sentence:** Can You Shoot an Apple   **Sentence:** BIG CATS like boxes too!
Morning Routine Brushing Teeth,   Thanksgiving Turkey Bowling | FACE   Off Your Head? (William Tell Archery
Taking a Bath, and Eating Breakfast!   OFF   Challenge)

Fig. 9.2   Example of YOVO-3M dataset

a validation set of 16,439 videos, and a test set of 16,554 videos. Participants need to submit the prediction results of the test set for the final ranking, and the test set labels are not disclosed. These videos are categorized into 240 predefined classes, including objects (such as aircraft, pizza, football) and human actions (such as waggle, high jump, riding).

The competition uses the F1 score as the evaluation metric.

9.3   Champion Solution

The architecture of the champion solution for this competition is shown in Fig. 9.3.

First, preprocess and augment the video data to obtain a dataset composed of image sequences; then, design pre-training tasks and conduct pre-training on the pre-training dataset; next, design experiments in the model stage to select a suitable backbone, use pre-trained parameters for initialization and fine-tuning; finally, integrate multiple models to further enhance the effect.

1.   Data Preprocessing

In the data preprocessing stage, first decode and sample the video, converting it into a series of image sequences composed of continuous frames. Then use center cropping and scaling to fix the input frame size to 256×256. Meanwhile, normalize the data using the mean and variance from ImageNet, converting the data distribution to a standard normal distribution. Finally, specify the data format and adjust the channel order to NCTHW.

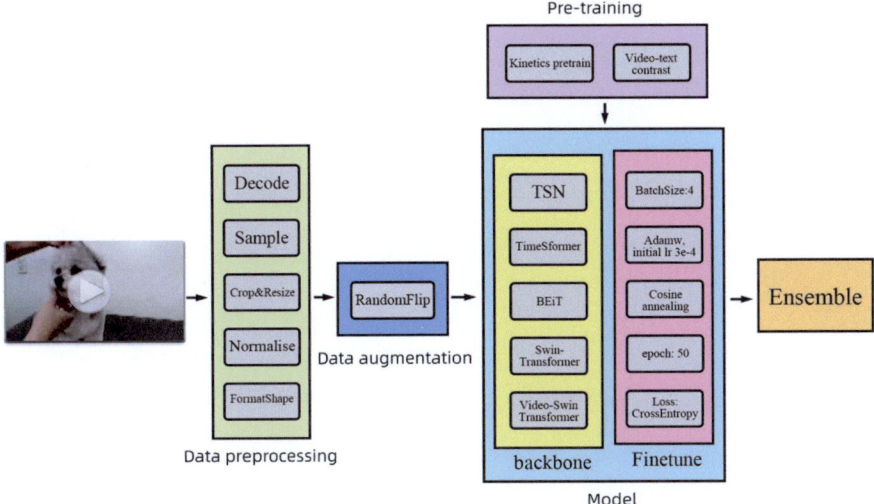

Fig. 9.3 Overall architecture of the champion solution

2. Data Augmentation

During the training phase, the RandomFlip strategy is used for data augmentation.

3. Pre-training

As shown in Fig. 9.4, a video-text contrastive learning pre-training mode similar to CLIP [1] is adopted in the pre-training stage.

The pre-training steps are as follows.

(1) Use two encoders to encode the text and video separately, obtaining video embeddings and text embeddings.
(2) Input the video embeddings into a projection head composed of three fully connected layers to obtain the classification probabilities of the video.
(3) Calculate the cross-entropy loss using the classification probabilities and the query.
(4) Project the two embeddings into the same dimensional space through linear transformation.
(5) Calculate the cosine similarity between each pair within the same batch to obtain the similarity matrix sim.
(6) Maximize the values of the diagonal elements of the sim matrix and minimize the values of other elements (using NCE loss).

The form of the NCE loss is as follows.

$$L_{\text{NCE}} = -\log \sum_{i=1}^{n} \frac{\exp(s_i, i/\tau)}{\sum_{j=1}^{n} \prod_{i \neq j} \exp(s_i, j/\tau)}$$

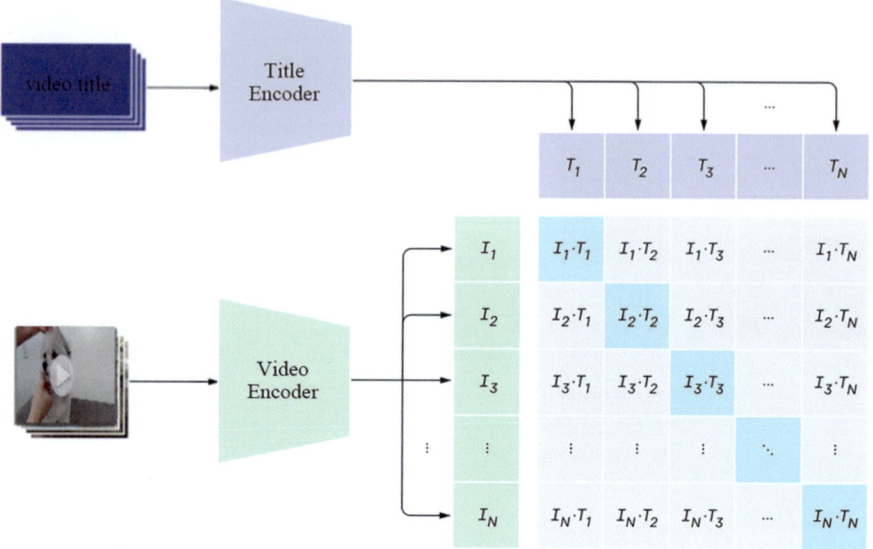

Fig. 9.4 Video-text contrastive learning diagram

Here it can be regarded as N with a temperature coefficient. N The categorical cross-entropy loss, where the diagonal represents the logits of the correct categories, and the other elements in the same row (column) are the logits of the incorrect categories.

The code for the pre-training part is as follows.

```
# Text and video encoding
I_f = video_encoder(I) #[n, d_i]
t_f = text_encoder(T) #[n, d_t]

#    Use    linear    projection    to    unify    dimen-
sions and perform L2 normalization
I_e = 12_normalize(np.dot(I_f, W_i), axis=1) #[n, d_l]
t_e = 12_normalize(np.dot(T_f, W_t), axis=1) #[n, d_l]

# Cosine similarity matrix: [n, n]
sim = cosine_similarity(I_e, t_e.T)

# Symmetric NCE loss
loss_i = NCE_loss(logits, labels, axis=0)
loss_t = NCE_loss(logits, labels, axis=1)
loss1 = (loss_i + loss_t)/2

# Query prediction
logits = cls(l_f)
```

```
loss2 = cross_entropy(logits, query)
loss = a*loss1 + b*loss2
```

In the specific implementation, the mmaction.model.recognizer.base.Base Recognizer class is inherited to create the VideoTextContrastRecognizer class, which wraps the two encoders and implements the training process. The specific implementation code is as follows.

```
@RECOGNIZERS.register_module()
class VideoTextContrastRecognizer(BaseRecognizer):
        def __init__(self, num_class=0, text_encoder_
path='roberta',
feature_dim=1024, **kwargs):
     # nitialize various parts of video-text contrast recogni-
tion
     super().__init__(**kwargs)
    self.con_loss = build_loss(dict(type='VideoTextContrast
Loss'))
        self.text_encoder = AutoModel.from_pretrained(text_
encoder_path)
    self.video_pool = nn.AdaptiveAvgPool3d(1)
    self.transform_video = nn.Linear(1024, 1024)
    self.transform_text = nn.Linear(768, 1024)
    if num_class != 0:
       self.is_classification = True
       self.classifier = nn.Sequential(
          nn.Linear(1024, 1024),
          nn.ReLU(),
          nn.Linear(1024, 1024),
          nn.ReLU(),
          nn.Linear(1024, num_class)
       )
                                self.cls_loss    =    build_
loss(dict(type='CrossEntropyLoss'))
    else:
       self.is_classification = False

            def    forward_train(self,imgs,text,attention_
mask,label=None,**kwargs):
    imgs = imgs.reshape((-1, ) + imgs.shape[2:])
    # Extract features of the video
    vf = self.extract_feat(imgs)
    # Encode the text and obtain the encoded text features
    tf = self.text_encoder(text).pooler_output
    # Transform video and text features into a new space
    vf = self.video_pool(vf)
    vf = vf.reshape((-1, 1024))
    transform_vf = self.transform_video(vf)
```

```
    transform_tf = self.transform_text(tf)
                        # Calculate the  cosine  simi-
larity matrix of video features and text features
        sim_matrix = torch.cosine_similarity(transform_
vf.unsqueeze(1), transform_tf.unsqueeze(0), dim=2)

    if self.is_classification:
        cls_score = self.classifier(vf)
    else:
        cls_score = None
    # Calculate loss and return
    return self.loss(sim_matrix, cls_score, label)
```

The implementation code for the NCE loss is as follows.

```
class VideoTextContrastLoss(nn.Module):
    def __init__(self, temperature=2, **kwargs):
        super().__init__(**kwargs)
        self.T = temperature
    def forward(self, sim_matrix):
        exp_sim = torch.exp(sim_matrix/self.T)
        row_sum = torch.sum(exp_sim, dim=0)
        col_sum = torch.sum(exp_sim, dim=1)
        diag = torch.diag(exp_sim)
        loss = (-torch.log(diag/row_sum) - torch.log(diag/col_
sum)) / 2
        return loss.mean()
```

4. Model Training

(1) Backbone Selection

The champion team tested five networks: Temporal Segment Network (TSN),[1] TimeSformer,[2] BEiT,[3] Swin Transformer,[4] and Video Swin Transformer (VST).[5] Table 9.1 shows the experimental results of different backbones on the validation set, including accuracy, parameter count, and FLOPs (floating point operations per second), with FLOPs indicating complexity.

Based on the experimental results, VST was ultimately used as the backbone. Additionally, since the competition does not impose restrictions on inference overhead, but only requiring the submission of inference results on the test set, the models

---

[1] The link is https://github.com/open-mmlab/mmaction2/tree/master/configs/recognition/tsn.

[2] The link is https://github.com/open-mmlab/mmaction2/tree/master/configs/recognition/timesf ormer.

[3] The link is https://github.com/microsoft/unilm/tree/master/beit.

[4] The link is https://github.com/microsoft/Swin-Transformer.

[5] The link is https://github.com/SwinTransformer/Video-Swin-Transformer.

Table 9.1 Experimental results of different backbones

Backbone	Accuracy	Parameter count	FLOPs
TSN-ResNet50	0.5214	24M	33.0G
TimeSformer	0.5440	121.4M	189.0G
BEiT-Large	0.5758	202.6M	358.3G
Swin Transformer-Large	0.5823	195.4M	272.3G
VST	0.6140	87.9M	60.6G

trained in the comparative experiments can also be used for inference and integrated into the final results.

(2) Training Details

The training process for each single model follows a similar approach. Here, the training process of the single model VST, which achieved the best results, is introduced. During training, the video encoder weights from the pre-training phase are first loaded. The Adam optimizer is used, cross-entropy is chosen as the loss function, the learning rate is initialized to $3e^{-4}$, and a cosine annealing learning rate adjustment strategy is adopted, training for a total of 50 epochs. The reference code is as follows.

```
optimizer         =         dict(type='AdamW',          lr=3e-
4, betas=(0.9, 0.999), weight_decay=0.05,
              paramwise_cfg=dict(custom_keys={'absolute_pos_
embed': dict(decay_mult=0.),
                                  'relative_position_bias_
table': dict(decay_mult=0.),
                        'norm': dict(decay_mult=0.),
                        backbone': dict(lr_mult=0.1)}))
# Learning Strategy
lr_config = dict(
   policy='CosineAnnealing',
   min_lr=0,
   warmup='linear',
   warmup_by_epoch=True,
   warmup_iters=2
)
total_epochs = 50
```

5. Model Integration

By sampling videos at different temporal sampling rates, training sets with different temporal resolutions are obtained, allowing for the training of different models. Ultimately, multiple models are integrated with equal weights (see Fig. 9.5). Due to resource and time constraints, the winning solution only conducted multi-resolution training and integration on the VST network.

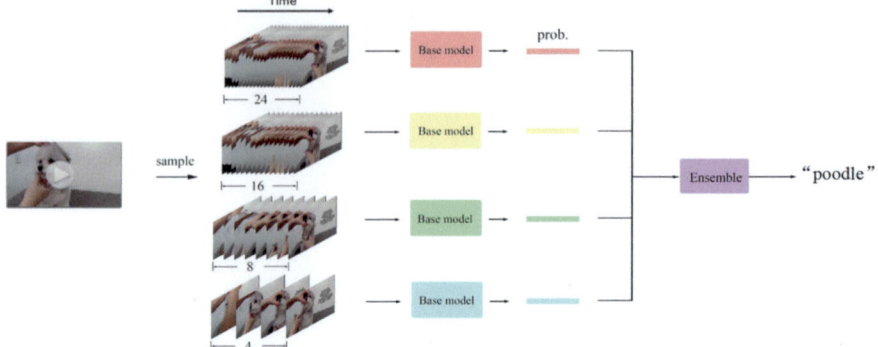

Fig. 9.5 Illustration of multi-temporal resolution integration

Table 9.2 Accuracy of single models with multi-temporal resolution and ensemble effect

Model classification	Accuracy
SR 4/12	0.6033
SR 8/6	0.6140
SR 16/3	0.5986
SR 24/2	0.6064
Ensemble	0.6193

Table 9.2 compares the experimental results of different sampling methods and ensemble learning, where SR x/y indicates sampling one frame every y frames, with a total of x frames sampled. It is evident that ensemble learning can significantly enhance the final performance.[6]

Reference

1. Radford, A., Kim, J. W., Hallacy, C., Ramesh, A., Goh, G., Agarwal, S., Sastry, G., Askell, A., Mishkin, P., Clark, J., Krueger, G., & Sutskever, I. (2021, February 26). Learning transferable visual models from natural language supervision [cs.CV, cs.LG]. arXiv https://doi.org/10.48550/arXiv.2103.00020

---

[6] The complete code can be found at https://github.com/DataBountyHunter/DataMiningCompetitionInAction.

Chapter 10
Reinforcement Learning: Theoretical Part

Reinforcement learning (RL) is a learning paradigm distinct from traditional supervised and unsupervised learning. Reinforcement learning is currently the closest approach to human learning, which involves learning through continuous trial and error. In reinforcement learning, an agent is typically defined, which can be a system that receives external inputs and then makes decisions. The agent can be constructed using deep neural networks or through dictionaries and tables. Agents constructed using deep neural networks are generally trained using deep reinforcement learning (DRL) algorithms. The part that provides external input to the agent is generally referred to as the environment. The environment can be a virtual game simulator (such as Honor of Kings, Go, Dota 2, etc.) or a real-world task (such as recommendation tasks, dialogue tasks, autonomous driving tasks, etc.).

Reinforcement learning involves the agent and the environment continuously interacting for learning. The reinforcement learning framework is shown in Fig. 10.1.

Figure 10.1 illustrates the interaction process between the agent and the environment in reinforcement learning. In this interaction process, the environment generates an environment state, which may not be fully received by the agent. The part that can be observed by the agent is called the observation. The agent makes corresponding actions based on the observation input and returns them to the environment. The environment, based on the actions given by the agent, returns new observations and rewards to the agent. The goal of the reinforcement learning agent is to maximize the cumulative reward over a period of time by making optimal actions. Unlike greedy algorithms, reinforcement learning agents do not focus solely on maximizing immediate rewards each time, aim to maximize cumulative rewards through a form similar to dynamic programming. Therefore, the current optimal action may not yield the maximum immediate reward in the next step, but it can achieve the global maximum reward.

Currently, the applications of reinforcement learning are very extensive, as shown in Fig. 10.2.

© Tsinghua University Press 2026
K. Xu, *Data Mining Competition Practices*,
https://doi.org/10.1007/978-981-95-3446-3_10

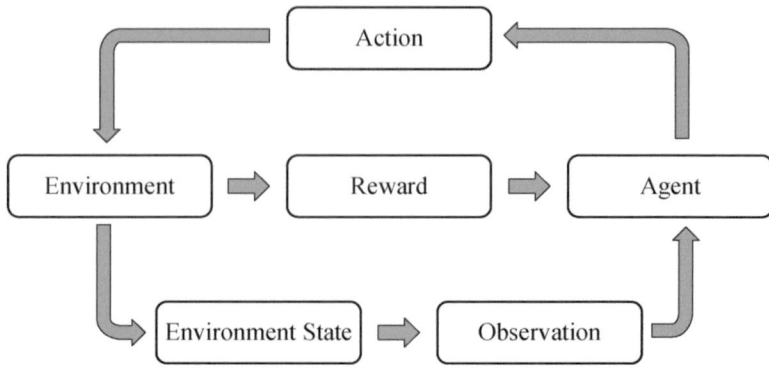

Fig. 10.1 Reinforcement learning framework

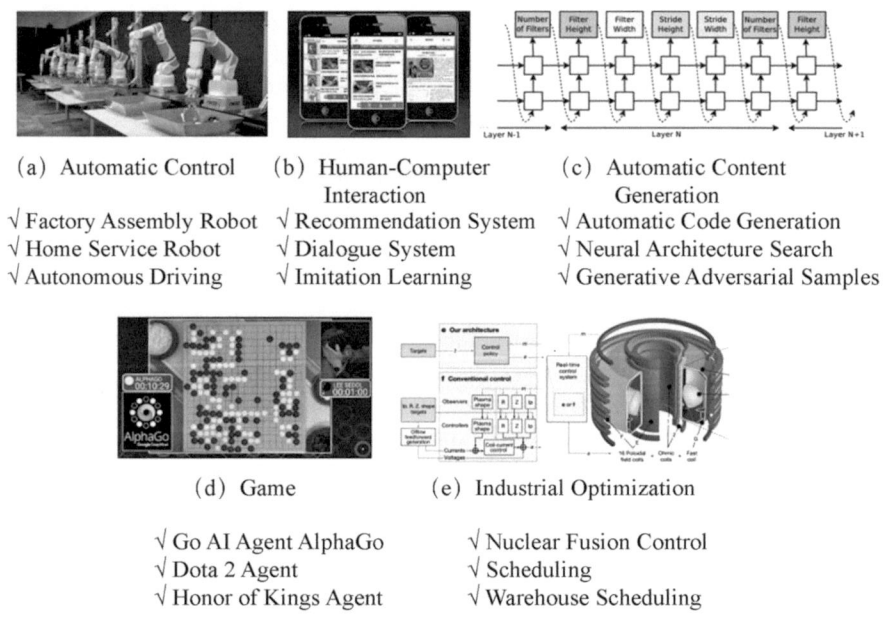

(a) Automatic Control   (b) Human-Computer   (c) Automatic Content
                                      Interaction   Generation

√ Factory Assembly Robot   √ Recommendation System   √ Automatic Code Generation
√ Home Service Robot   √ Dialogue System   √ Neural Architecture Search
√ Autonomous Driving   √ Imitation Learning   √ Generative Adversarial Samples

(d) Game   (e) Industrial Optimization

√ Go AI Agent AlphaGo   √ Nuclear Fusion Control
√ Dota 2 Agent   √ Scheduling
√ Honor of Kings Agent   √ Warehouse Scheduling

Fig. 10.2 Extensive applications of reinforcement learning

It can be seen that reinforcement learning plays an important role in fields such as game AI, nuclear fusion control, chip design, industrial scheduling, recommendation systems, quantitative trading, matrix multiplication acceleration, and robot control. Since the process of reinforcement learning does not require pre-collected labeled data and can train agents that exceed human performance through continuous interaction with the environment, reinforcement learning is also considered the learning paradigm closest to artificial general intelligence (AGI).

The design of reinforcement learning algorithms requires mastering the theoretical knowledge of reinforcement learning and combining it with extensive practical experience to formulate efficient solutions for different decision-making tasks. Next, we will introduce the design ideas for solving reinforcement learning tasks from three aspects: agent design, model design, and algorithm design.

10.1   Agent Design

First, we will introduce agent design in reinforcement learning. This section will cover observation input design, reward design, and action design.

10.1.1   Observation Input Design

A reinforcement learning agent needs to make decisions based on observations provided by the environment. Therefore, observation input is a key module for the agent to perceive the external environment. When designing observation input, on one hand, it is important to ensure the comprehensiveness of the information in the observation so that the agent can obtain enough information to make correct decisions; on the other hand, it is important to improve the effectiveness of the observation input. The so-called effectiveness mainly consists of two aspects: enhancing the comprehensiveness and effectiveness of the observation input.

1.   Enhancing the Comprehensiveness of Observation Input

In reinforcement learning tasks, the agent needs to make decisions based on observation input, and the comprehensiveness of the observation input is a decisive factor in whether the agent can make correct decisions. Taking Go as an example, if the agent can only observe a quarter of the board each time, no matter how efficient the subsequent reinforcement learning algorithm is, the agent cannot make the most correct decision. In the case of missing observation input information, it is difficult to train a powerful agent. Therefore, when designing observation input, it should include as much useful information as possible while avoiding redundant information. For example, in the multiplayer online battle arena (MOBA) game "Honor of Kings" (see Fig. 10.3), the agent's input should not only include information about the "hero" it controls (such as health, position, etc.) but also information about its teammates, opponents, and the entire match. This allows the trained agent to consider more global information and make more coordinated operations.

Additionally, for some decision-making tasks with incomplete information, the action taken by the agent in the previous step can also be used as the current observation input for the agent. This can help the agent generate more coherent actions both before and after. For example, in competitive fighting arcade games (such as King of Fighters, Naruto, One Piece: Fighting Path, etc.), the agent needs to complete

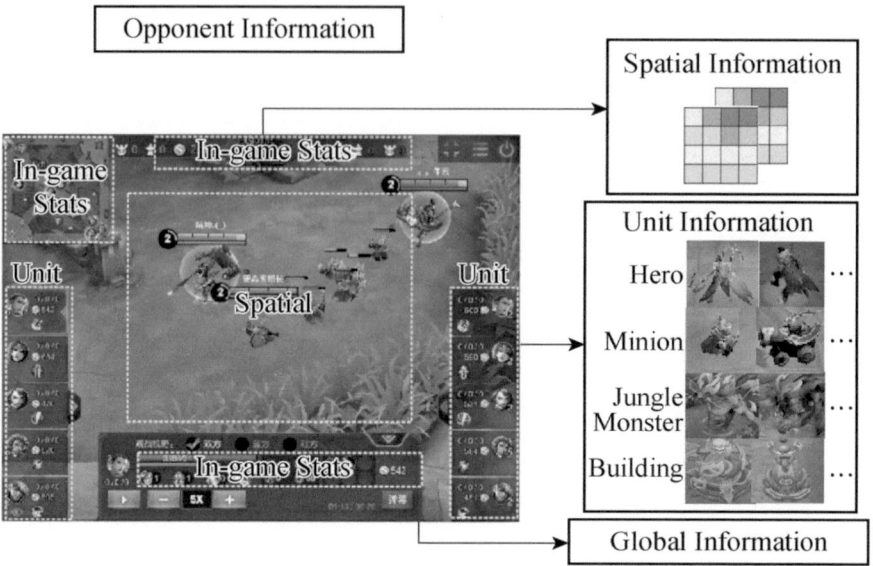

Fig. 10.3 Observation input of the honor of kings agent

specific skill releases and attack strategies through coherent actions. If the agent does not know what actions it has previously taken, it is difficult to perform combo actions. Therefore, in some decision-making tasks that are sensitive to the order of actions, including previous actions as part of the observation input is particularly important.

2. Enhancing the Effectiveness of Observation Input

Previously, we discussed how to enhance the comprehensiveness of observation input, but simply inputting all information to the agent without consideration may reduce the training efficiency of the agent. Therefore, enhancing the effectiveness of the agent's observation input is a direct method to improve the training efficiency and final performance of the agent. In this section, we mainly elaborate on two aspects: enhancing the relevance of the agent's observation input and enhancing the usability of the agent's observation input.

(1)   Enhancing the Relevance of the Agent's Observation Input

Relevance refers to the association between observation input and decision-making. For example, for an autonomous driving agent, the input information related to decision-making and planning operations should only includes vehicles and pedestrians on the road, while clouds in the sky and buildings on the roadside are irrelevant information that does not need to be input to the agent. Therefore, when designing the observation input for the agent, it is necessary to maximize the input of information related to decision-making as much as possible and minimize the input of information unrelated to decision-making. If the observation input contains too much irrelevant information, it will increase the difficulty for the agent to process and filter information, and may even overshadow useful information, leading to a decline in the agent's performance.

Additionally, different input information has varying impacts on the agent's decision-making. For example, when designing a short video recommendation agent, the visual information of the short video has a greater impact on the recommendation effect compared to the comments below it. Sometimes, the proportion of more relevant input information may be relatively small, and inputting all information together may detract from the agent's utilization of more important information. In this case, grouping input information, separating important information from relatively secondary information, allowing the neural network to process them separately and extract hidden features of different dimensions to balance the weight of different input information.

Judging the relevance of input information often requires analysis of specific tasks and dynamic adjustment of input based on training results. When there is redundancy in input information, it is necessary to reduce the input, which can be achieved by reducing the input dimension through neural network distillation. When additional input information is needed, the size of the neural network can be increased to allow the agent to add extra input without affecting performance. More details on the design of the agent's neural network will be introduced in Sect. 10.2.

(2) Enhancing the Usability of the Agent's Observation Input

Once it is determined which information needs to be used as input, the input format of the information must be determined. Different input formats pose different levels of difficulty for the agent to process. Generally, vector input is the simplest input format and is easier for neural networks to process. In addition to vector input, formats such as images can also be used, but they may increase the processing difficulty and training time for the agent. For example, in first-person shooter games, the current screen of the shooting game can be used as input for the agent, allowing the agent to obtain information such as remaining health, remaining ammunition, and enemy positions from the screen. Since the agent starts training randomly and does not have prior knowledge of game rules and items, it may take a very long game time for the agent to learn these basic concepts and apply them to decision-making. To enhance the usability of the agent's observation input, information such as health, ammunition, and position can be directly input as numerical vectors to the agent, allowing the agent to focus more on decision-making and planning rather than information extraction.

To enhance the usability of the agent's observation input, even vector input needs to be processed according to the actual situation. For example, in robot navigation tasks, due to the diversity of space and the uncertainty of space size, relative coordinates of the agent and other objects are generally used instead of absolute coordinates as input. This way, the agent will not be affected by the overall movement of the space. Additionally, for inputs such as numbers, since the numerical information of numbers does not inherently represent size and relative relationships, they are often represented using one-hot encoding (i.e., 0 is represented as [1, 0, 0], 1 as [0, 1, 0], 2 as [0, 0, 1]). In many board games, one-hot encoding is also used. One-hot encoding is used to represent the information of pieces in different positions.

Additionally, to enhance training stability, input vectors can be normalized: either through static normalization or dynamic normalization. In static normalization, the mean and variance of the input values are pre-calculated, and these fixed values are used to normalize the input during agent training. In dynamic normalization, the mean and variance of the input are calculated and updated in real-time during training, and the latest mean and variance are used for normalization each time. These methods can improve the usability of the agent's observation input, thereby reducing the difficulty of processing the observation input and ultimately enhancing the training efficiency of the agent.

10.1.2   Reward Design

The goal of a reinforcement learning agent is to maximize cumulative rewards, which can be the performance indicators we ultimately care about or a reward-punishment mechanism designed by humans. Generally, the denser the reward signal, the easier it is for the agent to learn, while relatively sparse reward signals increase the learning difficulty for the agent. Therefore, when designing a reward-punishment mechanism, it is desirable for the reward signal to be as dense as possible. For example, in a soccer game, if only the final number of goals is used as a reward, the agent would need a very long time of exploration to learn how to achieve the goal of scoring through a series of decisions. Therefore, additional reward signals can be added to help the agent achieve the final goal more quickly, such as adding rewards for ball possession and penalties for going out of bounds.

When designing reward signals, there are often different types of rewards, and it is necessary to balance the weights of these different rewards. Taking the soccer game as an example, the agent receives positive rewards in both scoring and ball possession situations, but these two rewards should not be equivalent. Since scoring is more important than ball possession, the reward for scoring should be designed to be greater than the reward for ball possession. Typically, different types of rewards can be directly summed to serve as the agent's final reward. However, directly summing different types of rewards means that only the same discount factor can be used, but in some cases, different rewards need different discount factors to achieve better results. In a soccer game, scoring rewards are more long-term decisions, while ball possession rewards are more short-term decisions, so the discount factor for scoring rewards should be set higher than that for ball possession rewards. To fit different rewards under different discount factors, a multi-head value network method can be used. The value network fits each type of reward separately, and the final value estimate is obtained by summing the different value outputs. It is worth noting that when designing or adding new rewards, the effectiveness and potential irrationality of the new rewards should be fully considered, otherwise, due to some special boundary conditions, the agent might find loopholes in the reward design, leading to unexpected situations.

10.1.3 Action Design

In reinforcement learning tasks, the agent needs to output actions to the environment. Generally, the form of actions is given by the environment, and the agent only needs to provide actions in the corresponding form. However, sometimes the original environment actions may not be reasonably designed, which can lead to low training efficiency for the agent. Therefore, different action forms can be designed for the agent, and then a conversion from agent actions to environment actions can be performed to ensure normal interaction between the agent and the environment. The key to designing agent actions is to use human knowledge to compress the original environment action space. Here, we will introduce several common techniques for action space compression.

1. Frameskip Technique

The Frameskip technique is widely used in many game agents, such as Atari games and ViZDoom games. Frameskip means that the action output by the agent at a certain moment will be executed for multiple steps, and the specific number of steps is a preset hyperparameter. By using the Frameskip technique, the number of decisions made by the agent can be reduced, thereby reducing the action search space of the agent over the entire trajectory cycle. When using the Frameskip technique, the agent's observations and rewards need to be adjusted accordingly. For example, the agent's reward should be the sum of rewards accumulated over the multi-step execution time, and the agent's observation can be that from the last step of the multi-step execution. Whether to use the Frameskip technique needs to be analyzed based on the specific task, as some tasks do not support repeated execution of actions multiple times.

2. Action Masking Technique

Action masking refers to dynamically blocking some unreasonable actions based on the agent's environmental state, allowing the agent to choose only from the remaining actions. For example, in autonomous driving, when a red light is detected, the agent should not make a forward action, and action masking can be used to block the forward action. The actual implementation of the action masking technique is to set the output probability of the actions to be masked to zero, so that the agent will not use the masked actions, regardless of whether it is through probability sampling or selecting the action with the highest probability.

3. Discretization of Continuous Actions

Agents with continuous actions are generally more difficult to train than those with discrete actions, so it is possible to consider converting continuous actions into discrete actions. For example, in a robot navigation task, the agent needs to provide the robot's steering angle. Since the steering angle is a continuous value from 0 to $360°$, allowing the agent to directly output this continuous value increases the difficulty of training the agent. Therefore, we can divide $0–360°$ into 12 equal parts, equivalent to $30°$ each. Then, when the agent makes a decision, it only needs to select one of these 12 parts and turn to the corresponding angle. In different tasks, continuous values can be discretized according to the actual needs of the task. When dividing 0–360° into 12 parts, the agent's steering precision is 30°. And if higher steering precision is required, it can be divided into 36 parts, making the steering precision 10°; if the steering precision requirement is not that high, it can be divided into 8 parts, so the agent will only set 45° as a unit of steering.

4. Grouped Action Output

In some decision-making tasks, it is necessary to output hierarchical actions. For example, in recommendation tasks, due to the large number of items to recommend, it is difficult to directly select a specific item from many items for recommendation. However, items can be classified. In this way, the agent can first decide which category of items to recommend and then select an item from this category. This greatly reduces the difficulty of the agent's decision-making. Additionally, when there are multiple types of actions, a separate output strategy can be adopted. For example, in a decision-making task, the agent needs to select one action from action set A and another from action set B. Action set A has 10 actions, and action set B also has 10 actions. The Cartesian product of action sets A and B is 100, and when using a neural network, this approach will cause the action output dimension to be too large, making the network difficult to train. Therefore, two different action output modules can be used, one for outputting actions from action set A and the other for outputting actions from action set B, so that the action output dimension of each module is 10, which helps to reduce the difficulty of training the neural network.

10.2 Model Design

Reinforcement learning agents can be implemented in various ways, with those based on deep neural networks being the most common form. In the process of training deep reinforcement learning, two types of neural networks are mainly involved: the policy network and the value network. The policy network outputs the actions the agent needs to execute based on the observation input, while the value network can be used to predict the cumulative expected return, thereby assisting in training the policy network. Generally, during the interaction between the agent and the environment, only the policy network is needed. However, in more complex reinforcement learning algorithms, such as those incorporating Monte Carlo search, the value network can also provide some assistance in the agent's decision-making. In this section, we will discuss how to design the policy network and the value network.

1. Policy Network Design

The design of a general policy network is similar to that of traditional supervised learning algorithms' neural networks. For vector inputs, a multi-layer perceptron (MLP) is usually used; for image inputs, a convolutional neural network is typically used; for natural language inputs, a Transformer network is often used. Additionally, if the agent's observation input is not complete information, the agent needs to have the ability to process historical information and possess memory. In this case, a recurrent neural network can be added to the policy network to handle historical information. Different top-level network structures are also required for different action outputs. If the agent needs to output discrete actions, a SoftMax layer is generally used to output the probabilities of different actions. If the agent needs to output continuous actions, a Gaussian distribution is generally used as the probability distribution of continuous actions, and the neural network needs to output the mean and variance of the Gaussian distribution. To reduce the difficulty of training, sometimes the variance of the Gaussian distribution is fixed, and the neural network only outputs its mean. To improve the efficiency of training the policy network, an auxiliary module for value prediction can also be added to the policy network, allowing it to participate in value prediction training. Additionally, in some multi-agent tasks, different agents may have observation inputs of the same dimension, allowing different agents to use the same policy network, and only different agent IDs need to be added to the observation input to distinguish different agents. By sharing the parameters of the policy network, both data utilization efficiency and training speed can be improved.

2. Value Network Design

The design of the value network is similar to that of the policy network and also requires using the corresponding basic network structure based on different observation inputs. Compared to the policy network, the value network can usually use global information as input. Since the reward signals of different tasks vary greatly in numerical size, a value normalization module is usually added to the value network to make value prediction training more stable. In some complex decision-making tasks, there may be different types of rewards, and a multi-head mechanism can be used in the value network to predict different rewards with different value prediction heads. Finally, the outputs of different prediction heads are summed to obtain the final total reward. Different value prediction heads can use the same underlying network structure, which can improve the efficiency of training data utilization.

10.3 Algorithm Design

For different decision-making tasks, training a reinforcement learning agent requires the use of appropriate reinforcement learning algorithms. This section will introduce some commonly used reinforcement learning algorithms. Since reinforcement learning algorithms involve numerous hyperparameters, this part will also introduce how to adjust hyperparameters. The process of reinforcement learning training is not very stable, so this section will also introduce some common training techniques. Additionally, in some tasks that involve adversarial elements, corresponding model performance evaluation methods are also needed.

10.3.1   Reinforcement Learning Algorithms

In recent years, researchers have proposed a variety of reinforcement learning algorithms. Generally, reinforcement learning algorithms can be divided into value-based reinforcement learning algorithms and policy-based reinforcement learning algorithms. value-based reinforcement learning algorithms only train a value network, and the agent selects the action with the maximum value for decision-making. The deep Q-learning network (DQN) algorithm proposed by DeepMind is a classic value-based reinforcement learning algorithm. Policy-based reinforcement learning algorithms require training a policy network, which can directly output the actions needed by the agent. A typical algorithm is the policy gradient algorithm. When the action space of the decision-making task is discrete, both types of algorithms can be used. However, when the action space of the decision-making task is continuous, policy-based reinforcement learning algorithms are generally used.

Additionally, reinforcement learning algorithms can also be distinguished based on the usage of training data. If the training data comes from previously trained agents, such algorithms are called off-policy reinforcement learning algorithms, such as the DQN algorithm and the A2C algorithm. If the training data comes from the current agent, such algorithms are called on-policy reinforcement learning algorithms, such as the TRPO and PPO algorithms. Generally speaking, if training data is difficult to obtain, off-policy reinforcement learning algorithms can be used, but they may cause a decline in agent performance due to data quality issues. When data is relatively abundant, on-policy reinforcement learning algorithms are usually used.

If a decision requires controlling multiple agents simultaneously, multi-agent reinforcement learning algorithms can be used, such as the QMIX and MAPPO algorithms. Additionally, due to the complexity of multi-agent tasks, issues such as communication between agents, neural network weight sharing, and decision consistency between agents may also arise, requiring targeted algorithm design.

Besides traditional reinforcement learning algorithms, there are many other specific reinforcement learning algorithms. For example, hierarchical reinforcement learning algorithms and enhanced exploration reinforcement learning algorithms for sparse reward problems, model-based reinforcement learning algorithms for board game tasks, offline reinforcement learning algorithms for offline data training, and adversarial training algorithms for multi-agent adversarial training. Thus, when facing different reinforcement learning tasks, it is necessary to draw on and design corresponding algorithms to solve the problem.

10.3.2   Hyperparameter Tuning

Numerous hyperparameters are involved in reinforcement learning training. We need to understand these hyperparameters to set more reasonable values. In reinforcement learning, important hyperparameters include the reward discount rate, learning rate, number of environment parallels, number of training updates, and the backpropagation length of recurrent neural networks. The reward discount rate is generally denoted by λ, and its main function is to balance the weight between short-term and long-term rewards. The value of the reward discount rate λ ranges from 0 to 1. The smaller the value of λ, the more the agent focuses on short-term rewards; λ The larger the value of λ, the more the agent focuses on long-term rewards. Generally speaking, when the rewards in the environment are more sparse, the value of the reward discount rate should be set higher to help the agent focus on more long-term strategies. Additionally, since the training process of reinforcement learning is more unstable compared to supervised learning, the learning rate is usually set lower to prevent inaccurate gradient updates from causing training crashes. The number of environment parallels is generally set as large as possible based on the number of CPUs and the size of the GPU memory, obtaining more sampling data through a larger number of environment parallels, thereby reducing estimation bias and improving the accuracy of gradient updates. Furthermore, the number of training updates is also an important factor affecting reinforcement learning training. When the number of training updates is too small, it may lead to low data utilization, resulting in inefficient training; when the number of training updates is too large, it may cause some online policy reinforcement learning algorithms to experience severe offline data bias. Additionally, for some incomplete information decision problems, recurrent neural networks are usually used to extract historical information. When training recurrent neural networks, the issue of setting the backpropagation length arises. If the backpropagation length is set too short, it may lead to inefficient training of the recurrent neural network; if the backpropagation length is set too long, it may result in a small batch size, leading to fewer data samples and unstable training. When manually adjusting training parameters, one can only analyze the specific reasons based on the training situation and then adjust the parameters accordingly. To reduce the complexity of manually adjusting parameters and the threshold for parameter adjustment, some hyperparameter automatic search tools can also be used, such as using wandb (see Fig. 10.4) for hyperparameter search and performance visualization. The detailed introduction on how to use wandb for hyperparameter search is not provided here.

10.3.3 Training Techniques

To improve the efficiency and stability of reinforcement learning training, some additional training techniques can be selectively added. Previously, techniques such as action masking, observation input normalization, value normalization, and the use of recurrent neural networks have been introduced. Here, we introduce some other commonly used techniques. For example, in the PPO algorithm, the generalized advantage estimation (GAE) technique can be used to reduce the estimation bias and variance of the value function. Below is a sample code of GAE implemented in Python.

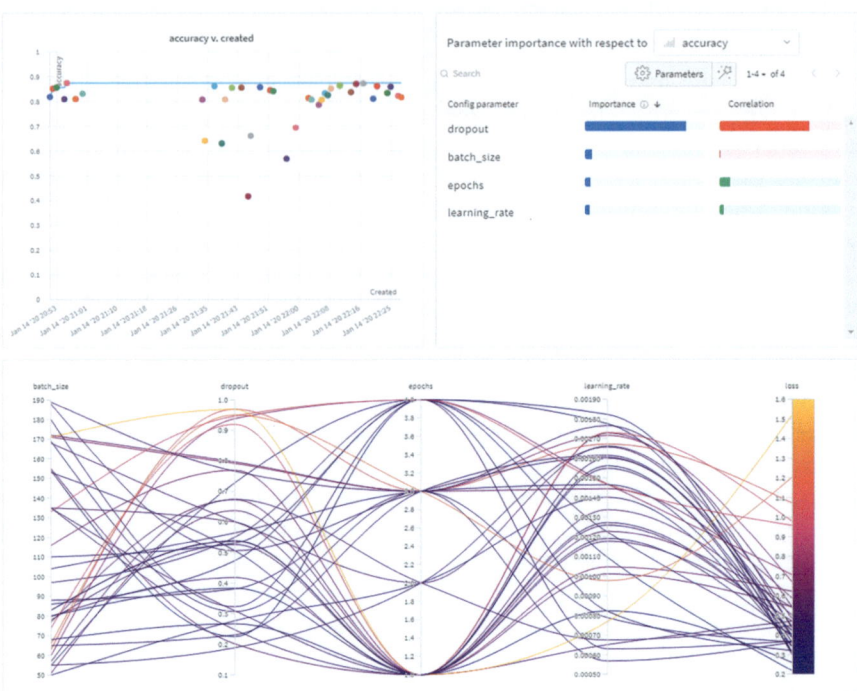

Fig. 10.4 wandb hyperparameter search interface

```
def discount_rewards(r, gamma=0.99, value_next=0.0):
    """

    Calculate the discounted sum of future rewards for updating
value estimates
      :param r: list of rewards
      :param gamma: discount factor
      :param value_next: T+1 value estimate for return calculation
      :return: discounted sum of future rewards as a list
      """

    discounted_r = np.zeros_like(r)
    running_add = value_next
    for t in reversed(range(0, r.size)):
        running_add = running_add * gamma + r[t]
        discounted_r[t] = running_add
    return discounted_r

def      get_gae(rewards,      value_estimates,      value_
next=0.0, gamma=0.99, lambda=0.95):
    """

            Calculate    generalized    advantage    estima-
tion for policy updates
      :param rewards: list of rewards from time step t to T
```

```
        :param value_next: value estimate at time step T+1
              :param   value_estimates:   list   of   value   esti-
   mates from time step t to T
        :param gamma: discount factor
        :param lambda: GAE weighting factor
        :return: list of advantage estimates from time step t to T
        """
        value_estimates = np.append(value_estimates, value_next)
         delta_t = rewards + gamma * value_estimates[1:] - value_
   estimates[:-1]
                        advantage     =     discount_rewards(r=delta_
   t, gamma=gamma * lambda)
        return advantage
```

The code analysis is as follows.

Here, rewards are the returns at each moment, value_estimates are the predicted values, value_next is the value at the next moment, gamma is the discount factor, and lambda is the weight factor of GAE.

Additionally, maximizing entropy can be used to enhance the exploration capability of agents during training. To increase the sampling volume and training speed, multi-machine distributed training can also be employed. When the overall data quality is not high, selectively discarding some low-quality data and retaining more high-quality data for training can allow the agent to focus more on the data that improves policy performance. Furthermore, some common training techniques in supervised learning can also be applied to reinforcement learning training. For instance, for observation inputs containing images, data augmentation can be performed by cropping and color perturbation, which can increase the training data samples and enhance the agent's robustness to observation input disturbances. Additionally, for neural network training, techniques to limit the size of gradient updates can be used to prevent the problem of gradient explosion. In complex real-world tasks, using offline data for model pre-training first, and then initializing with the pre-trained model for reinforcement learning training, is also an effective means to improve training efficiency. Due to the complexity of reinforcement learning training, the proficient use of training techniques is an important part of further enhancing agent performance based on existing performance. There are many training techniques in reinforcement learning that require continuous learning and development in practice.

10.3.4   Algorithm Performance Evaluation

The most direct way to evaluate the performance of reinforcement learning algorithms is to look at the rewards obtained by the agent. However, sometimes the reward size cannot reflect the final performance. For example, in a soccer game, the final performance of the agent needs to be evaluated through the win rate. Additionally, rewards are often just simple numerical values, sometimes a simple subtraction of multiple indicators, making it difficult to analyze various metrics of the agent from the rewards. For instance, in autonomous driving tasks, the agent's performance also needs to be evaluated from multiple aspects such as collision rate, number of sudden stops, and total driving distance. Moreover, for adversarial decision-making tasks, due to the presence of various opponents in the environment, it is impossible to directly evaluate an agent's overall strength using rewards. Therefore, in competitive games, specific ranking and scoring algorithms are usually used to evaluate the agent's performance, with commonly used algorithms including ELO and TrueSkill. Additionally, the training efficiency of the agent is also an important evaluation metric in algorithm research. In supervised learning, the number of update iterations when the training algorithm reaches convergence is usually used as the training efficiency metric. In contrast, reinforcement learning generally uses the number of interactions with the environment as the training efficiency metric. This is because reinforcement learning data is continuously generated, and generating training data requires consuming a large amount of hardware resources and sampling time, so the number of interactions with the environment is often more critical than the number of network parameter updates.

Chapter 11
Reinforcement Learning: Practical Part

This chapter, along with the competitive football game on the platform as a case study, introduces the specific application of reinforcement learning in practice.[1]

11.1 Competition Task

Football is one of the most popular sports in the world. This sport requires a balance between short-term control, passing, and high-level strategies, making it extremely challenging for AI to learn these strategies. The Google Football Game is a football game environment developed by Google for reinforcement learning algorithm research.[2] Figure 11.1 is a screenshot of the Google Football Game. The game task involves players from two sides, where each player needs to control all the team members of one side. Additionally, each player needs to submit their control code on the platform, which automatically conducts matches between different players, and then scores and ranks them based on the results of the matches.

11.2 Environment Introduction

This game involves two sides. In this game, each side will control 11 players from an 11-player team. The game rules are similar to real-world football rules, including offside, yellow cards, and red cards. The game is divided into two halves, each half lasting 45 minutes (1500 steps), totaling 3000 steps, and the game ends after 3000 steps. The kickoff at the start of each half is completed by different teams, but

---

[1] The link is http://www.jidiai.cn/env_detail?envid=34.

[2] The link is https://github.com/google-research/football.

K. Xu, *Data Mining Competition Practices*,
https://doi.org/10.1007/978-981-95-3446-3_11

Fig. 11.1 Screenshot of the Google football game

there is no exchange between the two sides (the match is completely symmetrical). Teams do not switch sides during the match. The left/right side is randomly assigned. The game simulator provides a dictionary format observation input, and users can construct their own vector input based on this dictionary format observation input. It mainly includes the position information of each player, the position information of the ball, and some match states on the field (such as score, yellow cards, offside information, etc.). Each player needs to design algorithms to control the 11 players on the field. Each player can execute one of 19 actions at each decision step.[3]

11.3 Evaluation Metrics

In this game, players score one point for each goal scored. The platform ranks players by randomly pairing them for matches and calculating the average score of the last 30 matches for different players. Figure 11.2 shows the scoring and ranking system of the JiDi platform.

---

[3] For specific meanings of the dictionary format observation input and agent actions, you can refer to https://github.com/google-research/football/blob/master/gfootball/doc/observation.md.

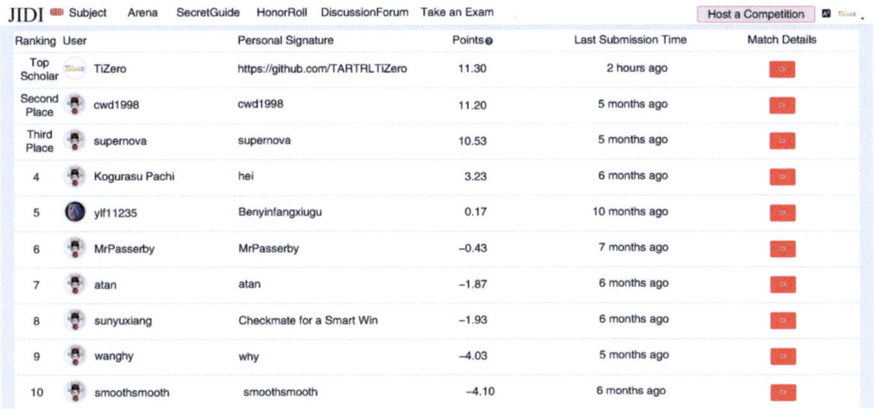

JIDI Subject Arena SecretGuide HonorRoll DiscussionForum Take an Exam Host a Competition

Ranking	User	Personal Signature	Points	Last Submission Time	Match Details
Top Scholar	TiZero	https://github.com/TARTRLTiZero	11.30	2 hours ago	
Second Place	cwd1998	cwd1998	11.20	5 months ago	
Third Place	supernova	supernova	10.53	5 months ago	
4	Kogurasu Pachi	hei	3.23	6 months ago	
5	ylf11235	Benyinfangxiugu	0.17	10 months ago	
6	MrPasserby	MrPasserby	−0.43	7 months ago	
7	atan	atan	−1.87	6 months ago	
8	sunyuxiang	Checkmate for a Smart Win	−1.93	6 months ago	
9	wanghy	why	−4.03	5 months ago	
10	smoothsmooth	smoothsmooth	−4.10	6 months ago	

Fig. 11.2 Scoring and ranking system of the JiDi platform

11.4 Champion Solution

This section mainly introduces the reinforcement learning agent TiZero designed for the Google Football Game. Through the design of the TiZero agent, readers can understand how to solve a reinforcement learning task from scratch. The complete code for TiZero can be accessed via the link https://github.com/TARTRL/TiZero.

1. Observation Input Design

First, design the observation input for the agent. For the policy network, the observation input is mainly divided into six parts: the information of the current ball-controlling player, the player number, the football information, the teammate information, the opponent information, and the current match information. For the value network, its input is divided into five parts: the football information, the ball holder information, the own team player information, the opponent team player information, and the current match information. The observation input for the policy network and the value network is shown in Tables 11.1 and 11.2.

Table 11.1 Policy network observation input

Observation input type	Dimension
Information of the current ball-controlling player	87
Player number	11
Football information	57
Teammate information	88
Current match information	88
Information of the current ball-controlling player	9

Table 11.2 Value network observation input

Observation input type	Dimension
Football information	23
Ball holder information	12
Own team player information	88
Opponent player information	88
Current match information	9

2.   Network Structure Design

TiZero has designed one policy network and one value network for the agent. The policy network uses six different sets of fully connected layers to extract observation input information from different parts. ReLU activation layers and layer normalization are used in the fully connected layers. Then, TiZero uses LSTM to record historical information. When the policy network outputs actions, TiZero masks illegal actions using an action mask. The structure of the policy network is depicted in Fig. 11.3.

Additionally, the value network uses five different sets of fully connected layers to extract observation input information from different parts. ReLU activation layers and layer normalization are used in the fully connected layers. Then, TiZero uses LSTM to record historical information. The value network is trained by minimizing the mean-squared error loss function. The structure of the value network is depicted in Fig. 11.4.

3.   Reward Design

The Google Football Game provides a basic reward mechanism, where the own team scores a goal and receives a reward of + 1, while the opponent team scores a goal and receives a reward of − 1. Since the number of goals in the Google Football Game is

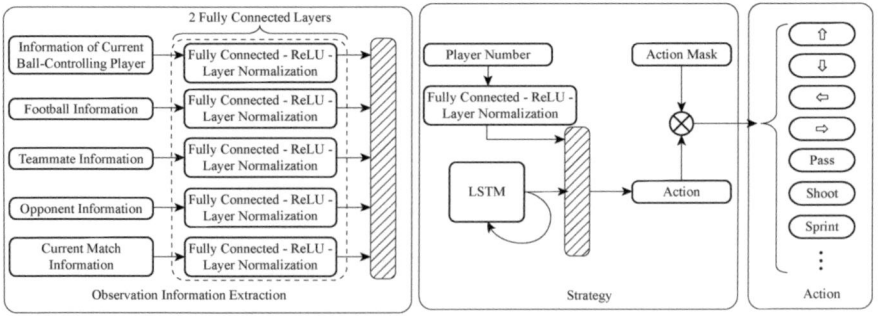

Fig. 11.3 Structure of the policy network

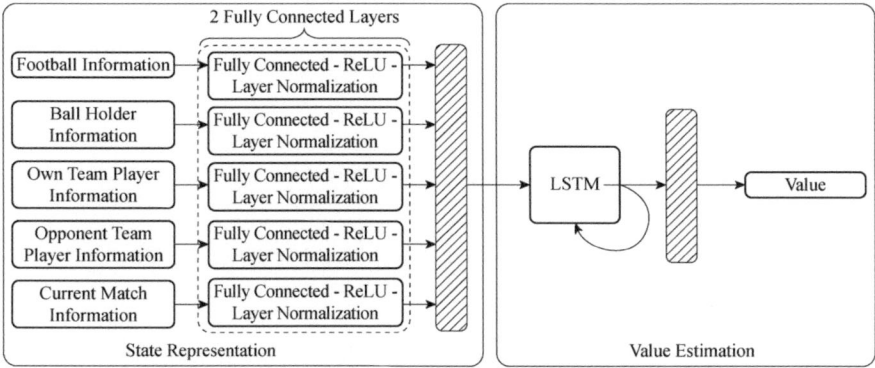

Fig. 11.4 Structure of the value network

not many, the reward signal is very sparse, which causes great difficulty in training the agent. To improve the training efficiency of the agent, TiZero also customizes some heuristic rewards. These additional rewards are listed as follows.

- Ball Possession Reward. The team in possession of the ball will receive a reward of + 0.0001.
- Passing Reward. Successful passes before scoring a goal will receive a reward of + 0.05.
- Clustering Penalty. If a team's players cluster together, they will receive a reward of − 0.001.
- Out-of-Bounds Penalty. When an agent goes out of bounds, it will receive a reward of − 0.001.

Since TiZero is trained through self-play, if all rewards meet the requirements of a zero-sum game, that is, when one team gains a certain reward, the opponent team loses the same reward, achieving a total reward sum of zero for both teams.

4. Action Design

Since the competition only allows the use of 19 specified actions, participants cannot arbitrarily modify the actions of the agent. However, action output optimization can be achieved through the action masking technique introduced in Sect. 10.1.3, thereby reducing the action search space and improving the training efficiency of the agent. For example, when the team that the agent controls has possession of the ball, tackling should be prohibited. Additionally, when a player is relatively far from the ball, passing and shooting should be prohibited. More details on the design of action masks can be found in the TiZero codebase.

5. Reinforcement Learning Algorithm

TiZero has designed a multi-agent reinforcement learning algorithm called Joint-Ratio Policy Optimization (JRPO). TiZero uses a value network to generate consistent values for all agents $V_{\text{total}}(s_t)$, and then calculates the overall advantage value

$A_{total}(s_t, a_t)$ using the GAE algorithm, which can be abbreviated as $\hat{A}_t$. TiZero uses this advantage value to guide the improvement of each agent's policy. TiZero employs the following policy decomposition form:

$$\pi_\theta(a_t|o_{1:t}) \approx \prod_{i=1}^{n} \pi_\theta^i(a_t^i|o_{1:t}^i)$$

Then, the following objective function is used as the training loss function for the policy network:

$$L^{CLIP}(\theta) = \hat{E}_t\big[\min\big(r_t(\theta)\hat{A}_t, \text{clip}(r_t(\theta), 1-\epsilon, 1+\epsilon)\hat{A}_t\big)\big]$$

where

$$r_t(\theta) = \frac{\pi_\theta(u_t|o_{1:t})}{\pi_{\theta_{old}}(u_t|o_{1:t})} = \prod_{i=1}^{n} \frac{\pi_\theta^i(u_t^i|o_{1:t}^i)}{\pi_{\theta_{old}}^i(u_t^i|o_{1:t}^i)}$$

And $\hat{E}_t[\ldots]$ represents the expectation over the sampled data.

6. Model Performance Evaluation

TiZero is trained through self-play, and the rewards are zero-sum; therefore, the performance of the agent cannot be judged solely by observing cumulative rewards. Agents trained through self-play are typically evaluated using TrueSkill. Figure 11.5 shows a comparison of TrueSkill scores for agents trained by different methods in the Google Football game. It can be seen that TiZero achieved a score significantly higher than other agents after more than forty days of training.

Although TrueSkill can conveniently evaluate the overall performance of agents, it does not directly reveal the advantages and behavioral characteristics of agents to humans. Therefore, comparing other objective indicators can help demonstrate the behavioral characteristics of different agents. Table 11.3 shows a comparison of different agents on various performance indicators (such as the number of assists, number of passes, pass success rate, etc.). As shown, the agent controlled by TiZero has the highest number of assists.

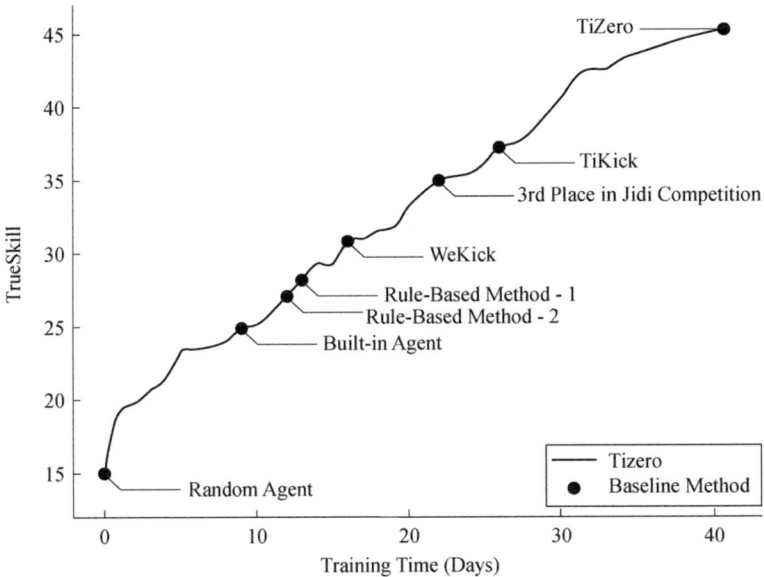

Fig. 11.5 Comparison of TrueSkill scores for agents trained by different methods in the game

Table 11.3 Comparison of different agents on various performance indicators

Indicator	TiZero	TiKick	WeKick	Third place in competition	Built-in agent	Rule-based method-1	Rule-based method-2
Assists	1.30 (1.02)	0.61 (0.79)	0.20 (0.47)	0.35 (0.62)	0.20 (0.55)	0.28 (0.59)	0.22 (0.53)
Passes	19.2 (3.44)	6.99 (2.71)	5.33 (2.44)	3.96 (2.33)	11.5 (4.63)	7.28 (2.77)	7.50 (3.12)
Pass success rate	0.73 (0.07)	0.65 (0.17)	0.53 (0.18)	0.44 (0.19)	0.66 (0.12)	0.64 (0.17)	0.63 (0.19)
Goals scored	3.42 (1.69)	1.79 (1.41)	0.88 (0.88)	1.43 (1.34)	0.52 (0.91)	0.73 (0.69)	0.64 (0.82)
Goal difference	2.27 (1.93)	0.71 (2.08)	− 0.47 (1.68)	− 0.02 (2.14)	− 1.06 (1.93)	− 0.60 (1.03)	− 0.71 (1.45)
Draw rate/%	8.50	22.2	29.0	23.2	24.8	28.7	27.8
Loss rate/%	6.50	23.5	44.2	33.8	59.6	48.2	49.5
Win rate/%	85.0	54.3	26.8	43.0	15.6	23.1	22.7
TrueSkill	45.2	37.2	30.9	35.0	24.9	28.2	27.1